Topics in Energy

Edited by M. Grenon, S. Srinivasan, G. Woite

G. Beckmann
P. V. Gilli

Thermal Energy Storage

Basics – Design – Applications to Power Generation and Heat Supply

Springer-Verlag Wien NewYork

Dipl.-Ing. Dr. techn. Georg Beckmann
Deputy Head, Industrial Plant Department, Waagner-Biro AG, Vienna, Austria

o. Univ.-Prof. Dipl.-Ing. Dr. techn. Paul Viktor Gilli
CEng, M. I. Mech. E., Mem. ASME, ASHRAE
Head, Institute of Thermal Engineering, Graz University of Technology, Austria

Printed in Austria

With 176 Figures

Library of Congress Cataloging in Publication Data. Beckmann, G. Thermal energy storage. (Topics in energy, ISSN 0723-4570.) 1. Heat storage. 2. Heat storage devices. 3. Electric power-plants. 4. Heating plants. I. Gilli, Paul Viktor. II. Title. III. Series. TJ260.B424. 1984. 621.402'8.83-20101

ISSN 0723-4570
ISBN 3-211-81764-6 Springer-Verlag Wien-New York
ISBN 0-387-81764-6 Springer-Verlag New York-Wien

Preface

This book deals with the basics of thermal energy storage and its applications to energy conservation and oil substitution in industry, in space heating (including cogeneration of heat and power), in peak load shaving and in vehicle drives.

The number of proposed and built thermal storage systems and their application have increased greatly during the last years. This is because of the increase of nuclear power, the utilization of solar energy, the availability of new vessel technologies and the need for oil substitution in power generation, in industry as well as in the home and in vehicles. The importance of thermal energy storage has increased accordingly.

We gratefully acknowledge the kind assistance of the companies that provided us with the illustrative material and data necessary for the writing of this book.

Vienna and Graz, December 1983 **G. Beckmann** and **P. V. Gilli**

Contents

1. Introduction

1.1. The Basic Task of Energy Storage

Supply systems for secondary energy consist of a primary energy source, an energy conversion system, and secondary energy consumers. There will be discrepancies—time-wise and sometimes local—between energy supply and demand. Overcoming these discrepancies is the basic task of energy storage [1.1–1.32]*, a task which has now become a topic in its own right, with separate scientific symposia [1.1–1.16, 1.30, 1.32].

If such discrepancies are caused by (more or less sudden) changes of demand, it is the peak load problem that may be solved, or at least alleviated, by means of energy storage. An energy storage plant may, on the one hand, exhibit lower specific investment costs than a peak load plant; and, on the other hand, it may lower fuel cost despite some storage losses, since excess energy from base load plants with low fuel cost may be used to charge the store.

If discrepancies between supply and demand are caused by the kind of primary energy source and energy converter, then the task of the storage plant is to smooth production not only by shaving production peaks but also by filling production valleys (bottoming). It is typical of some primary energy sources that their power is subject to variations such as systematic (periodic) variations in solar and tidal power plants and stochastic variations in wind, hydro and solar power plants.

Further tasks for energy storage include:

- instantaneous reserve in case of sudden outages of plants, particularly for bridging the gap until reserve plants are started up;
- control or buffer storage at high rates of load change, which will enable the energy conversion plant to cover the load with low gradients of load change;
- storage at or near the consumer, so as to reduce load peaks and plant cost not only at the energy conversion stage but also of the distribution system.

Fig. 1.1 shows the typical elements of secondary energy supply. In a heat supply system (A), energy storage will be thermal; it may be located at the heating plant itself, at the end of a long-distance pipe (before the load distribution

* References are quoted at the end of the book.

system), or at the consumer. The piping itself will be an energy storage system of considerable capacity. In an electric power supply system (B), energy storage may be within the power station (integrated storage); or it may be somewhere connected to the grid, or even at the consumer (after meter). The electric grid itself has practically no storage capacity.

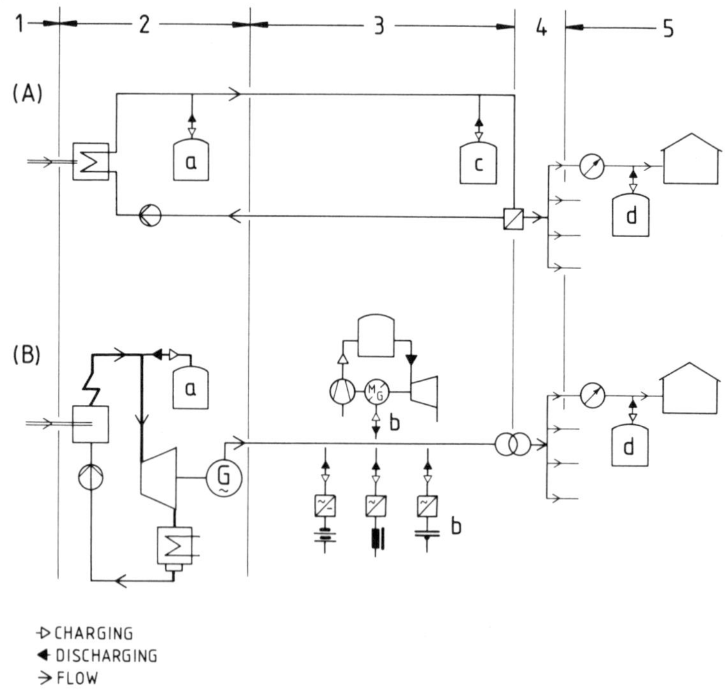

Fig. 1.1. Elements of secondary energy supply systems
A heat supply system, *B* electric power supply system, *1* primary energy source, *2* energy conversion, *3* secondary energy transport, *4* secondary energy distribution, *5* secondary energy consumer, *a* integrated storage, *b* storage in the grid, *c* storage before distribution, *d* storage after meter

Integrated storage may be thermal or mechanical; storage in the grid may be pumped thermal, pneumatic or hydraulic storage or direct storage such as electrochemical or electromagnetical, or in electric condensers. Storage at the customer may be thermal, electrochemical or mechanical.

1.2. The Economic Parameters of Energy Storage Plants

A full assessment of an energy storage plant includes not only the economic comparison of several possible storage and peaking systems but also an optimization of the size of one or more storage systems in an energy supply system. Sometimes, reasons other than economic ones (flexibility, availability, operational advantages) will be decisive for the installation of storage systems.

It should be noted that the optimization of a storage plant will need even more care and more knowledge of the demand and its time-dependence than the optimization of a pure conversion plant without storage. With storage plants not only the discharge power but also the storage capacity and the charge power are parameters that must be selected. Although typical cases may be treated in a general way, the number of parameters is such that each investment decision has to be considered on its own merits.

Installation costs of a storage plant consist of a part which depends on capacity and a part which depends on (discharge and charge) power:

(a) Capacity-dependent cost:

- storage vessel with liner, thermal insulation, support, etc.,
- storage medium;

(b) Discharge-power-dependent cost:

- discharge equipment within as well as external to the storage vessel,
- discharge piping,
- expansion engine and generator (or enlargement of main engine),
- heat disposal system (or enlargement of main system);

(c) Charge-power-dependent cost:

- charge equipment within as well as external to the storage vessel,
- charge piping,
- compression engine and motor (pumped storage only).

The remaining cost items such as buildings, foundations, control, installation, commissioning, etc. cannot easily be allocated to the above categories. Usually (b) and (c) are added up and referred to the discharge power, thus yielding the investment cost and specific investment cost

$$C_{st} = N_d c_{pow} + E_{st} c_{cap} \tag{1.1}$$

$$c_{st} = c_{pow} + d c_{cap} \tag{1.2}$$

where

C_{st}	installation cost of storage plant ($)
N_d	discharge power (kW)
c_{pow}	discharge-power-dependent part ($/kW)
E_{st}	storage capacity (kWh)
c_{cap}	capacity-dependent part ($/kWh)
$c_{st} = C_{st}/N_d$	specific investment cost ($/kW)
$d = E_{st}/N_d$	full load discharge duration (h) or "discharge hours".

In Table 1.1 specific investment cost of several types of storage plants for electric peak load coverage are given according to Messerschmidt-Bölkow-Blom (MBB, [6.20]). These data are plotted in Fig. 1.2 against discharge hours d. The diagram shows that thermal storage is more suitable for short duration (a few hours, for diurnal and possibly weekly storage), whereas hydraulic pumped storage (if proper sites are available) is more suitable for long durations d (durations of more than 12, up to several hundred hours, for weekly and seasonal storage).

Table 1.1. *Technical and cost characteristics of energy storage systems for peak load coverage*

	Fuel or Charging Energy	Specific Investment Cost[1] c_{st}		Turnaround Efficiency
		Power-dependent c_{pow} DM/kW	Capacity-dependent c_{cap} DM/kWh	
Gas Turbine (with Fuel Storage)	Oil	400	0.1	0.26–0.30[2]
Compressed Air Energy Storage with Combustion; CAES	Off-peak power, Oil, Gas	240	25	0.55[3]
Pneumatic Pumped Storage (without Combustion)	Off-peak power	300–380	70–150	0.65–0.7
Hydraulic Pumped Storage	Off-peak power	550	25	0.5–0.75
Advanced Batteries	Off-peak power	190–665	57–170	0.7
Flywheel	Off-peak power	450–500	100–150	0.7
Thermal Energy Storage (Hot Water)	Off-peak power	215–285	70–100	0.7–0.95

[1] DM, 1977 [6.20]; 1 DM (1977) = 0.44 \$ (1977).
[2] Thermal efficiency.
[3] For definition see Section 7.3.2.

The fuel cost of the storage plant is

$$b_{st} = s_c/\eta_{st} \tag{1.3}$$

b_{st} fuel cost (\$/kWh)
s_c cost of charge electric energy (\$/kWh)
η_{st} storage efficiency (turnaround efficiency).

The storage efficiency η_{st} (turnaround efficiency) is defined as the ratio of (electrical) discharge energy, E_d, to (electrical) charge energy supplied or produced less, E_c. For one cycle

$$\eta_{st} = E_d/E_c = \frac{E_d}{E_d + L} = \frac{1}{1 + L/E_d} \tag{1.4}$$

L losses per cycle (kWh).

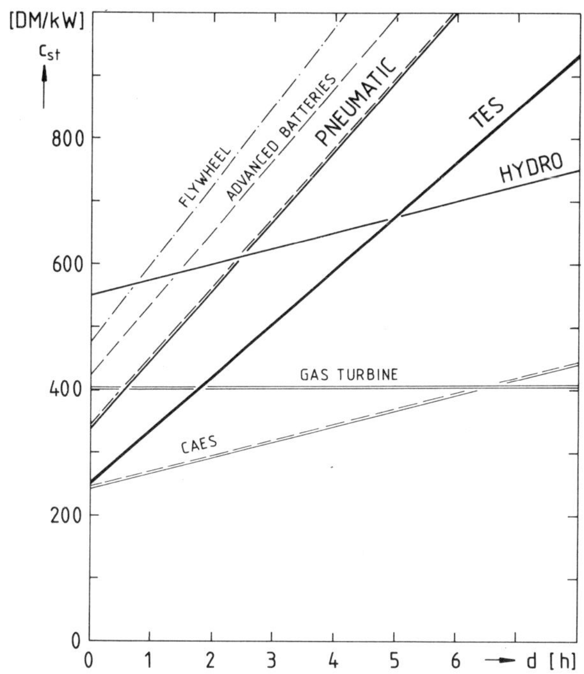

Fig. 1.2. Specific investment cost c_{st} of peak load storage plants versus full load discharge duration d (DM, 1977 [6.20])
CAES compressed air energy storage, *HYDRO* pumped hydraulic storage, *TES* thermal energy storage (pressurized hot water)

Typical values of η_{st} and other data are also given in Table 1.1. Thermal energy storage shows rather high values of η_{st} when compared to systems with a larger number of energy conversions.

The foregoing also applies to heat supply systems, with the proviso that only thermal storage is suitable, and no charge and discharge engines are required.

1.3. Optimization

1.3.1. Energy Supply Systems with One Energy Converter

An energy supply system with one type of energy converter is the simplest case. A boiler supplying a shop with steam or a building with hot water is an example of this (Fig. 1.3). Large fluctuations of steam demand are typical for consumers with a small steam network.

Part 1 of Fig. 1.4 shows the daily steam demand of an industrial plant (pulp) with distinct peaks. Without storage, the rated boiler power N_k would have to be equal to $N_{\max}$. However, with storage, N_k will be lower, and the more so the shorter the peaks and the higher the storage capacity. Optimization con-

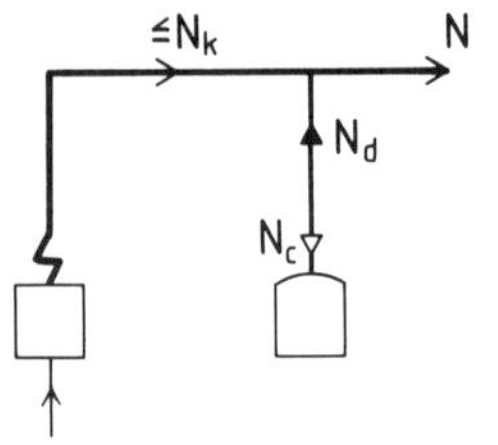

Fig. 1.3. Heat supply system with boiler and storage

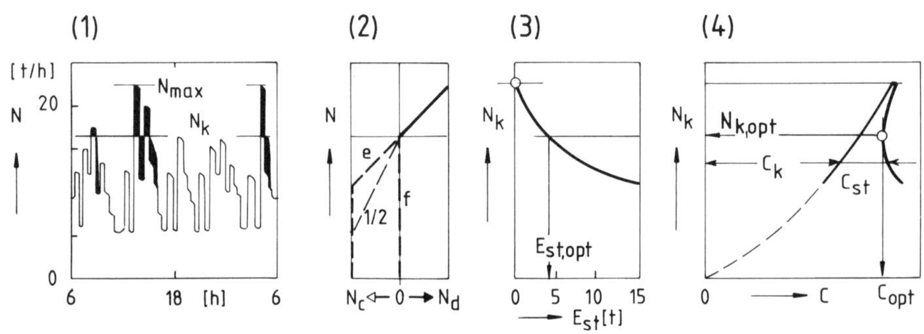

Fig. 1.4. Optimization of storage capacity
1 daily steam demand N, *2* operation of storage (*f* store fully charged, 1/2 store half charged, *e* store empty), *3* rated boiler power N_k versus storage capacity E_{st}, *4* installation cost of boiler (C_k) and storage (C_{st})

sists primarily of comparing investment cost with and without storage. Fuel cost will not be much different: The storage losses (mainly insulation losses) will increase, the smaller boiler with better utilization factor will decrease fuel consumption. It is assumed that the effects cancel each other out, i.e. $\eta_{st} = 1$. However, if this should not be the case, then the more complicated assessment according to Section 1.3.2 should be used.

Part 2 of Fig. 1.4 shows operation of the storage system. Above a certain load, discharge takes place; and charging takes place below a certain demand with consideration of the charge condition.

Part 3 shows on the abscissa storage capacity E_{st} necessary for this peak lopping. A small storage capacity already leads to a strong reduction of the necessary rated boiler power. However, the curve soon levels off, for a full boiler load equalization will normally not be cost-effective.

Part 4 of Fig. 1.4 presents the system cost which consists of boiler cost C_k and store cost C_{st}. The resulting curve yields the cost minimum C_{opt} at a certain rated boiler power $N_{k,\text{opt}}$ and storage capacity $E_{st,\text{opt}}$.

1.3.2. Multi-source Energy Supply Systems

This case may exist in an electric grid with power plants fed by various fuels, or in a district heating system that is based on combined heat and power generation (CHP), but whose peak loads are covered by heating boilers without power generation. If there are only two fuels, the cheaper one will be used for base load, and the more expensive one for peak load. For electric power generation, the base load might be covered by coal, nuclear power and/or hydro (river) plants, while the peak load might be covered by gas, oil or coal plants. For solar heating plants, solar energy will take the "base load"; that is, it will be used wherever possible, and conventional heating will take peak load.

The storage system will take up excess energy from the base load plant, store it and then will unload it again when demand exceeds the power of the base load plants. It will, therefore, save valuable peak load fuel. If it is possible (and predictable with some degree of certainty) to use the storage plant during the absolute maximum peak load, the necessary peak power is lower and the peaking plant cost is reduced, as described in Section 1.3.1. In actual operation near the maximum peak power, there will be a dilemma: For fuel saving reasons, the store should be discharged whenever the base load is exceeded; for covering the maximum load with a high degree of probability, the store should be kept fully loaded for as long as possible. Obviously, detailed knowledge of the demand curves (i.e., load prediction) is necessary for optimum operation. It is also necessary for economic assessment and optimization [1.29].

From a grid analysis, the annual number of discharges and the peak load substitution can be assessed as functions of storage capacity E_{st}. Strictly speaking, the procedure would have to be carried out for each day of the year and would require computer calculations. However, by making use of the daily, weekly, and seasonal periodicity, a few typical shapes of the daily load curve may be sufficient to achieve the necessary degree of exactness.

A daily load diagram typical for Central Europe is shown in Fig. 1.5 for the two extreme cases (mid-winter and mid-summer) and the medium case (spring and fall equinox). In mid-winter, the load reaches the highest value (N_{max}) for

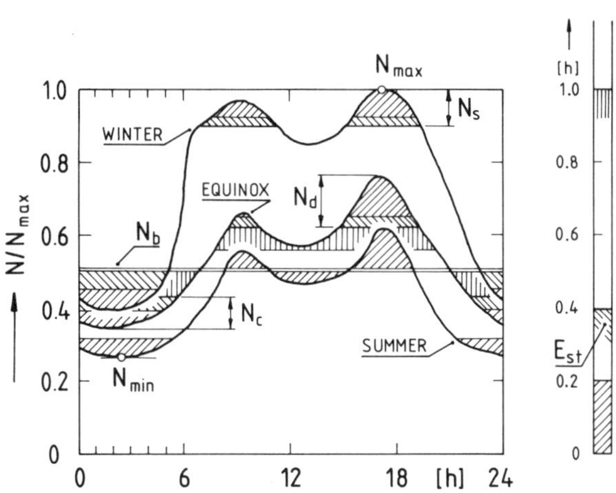

Fig. 1.5. Typical load diagram for Central Europe [1.29]

the entire year; in mid-summer, it reaches the lowest value (N_{min}). Spring and fall equinox are in between. Peaks occur in the morning and in the late afternoon.

The available power of the base load stations N_b is indicated in the diagram. It may be construed as being the installed capacity or, more correctly, a somewhat reduced capacity due to availability limitations. For annual load profiles such as the ones in the diagram, it will be different for winter and summer conditions: Maintenance of fossil-fired stations as well as maintenance and refueling of nuclear stations will preferably be done during the summer (low demand) season, with the consequence of a considerably lower available power during that period. It may also be preferable to close some base load power stations deliberately down during the low-load season and to run the remaining ones with a higher load factor employing the storage system. In this way, low part-load efficiency and/or fatigue due to thermal cycling is avoided. However, for the sake of simplicity only one line of available power is shown in Fig. 1.5.

What are the savings which result from the introduction of storage systems into a grid?

- During the winter peak, thermal storage saves installation cost as well as fuel cost due to a reduced requirement and reduced operation of peaking plants. Utilization of the storage plants will be limited by the available charging energy at the low-demand hours during the night.
- In the case of mid-summer, relatively little peak energy (above the line of available base load power N_b) will be required. Thermal storage saves fuel cost. Utilization of the storage plant will be limited by the peak energy required. If, during the low-demand season, the reduced availability of base load plant due to maintenance and refueling is taken into account, the saving of fuel cost is increased.
- In the case of equinox, the utilization of storage plants may or may not be limited by either the available charging energy (as for the mid-winter condition) or by the required peaking energy (as for the mid-summer condition).

In Fig. 1.5 the possible storage utilization in case of 100% storage efficiency (indicated by different shading in the three cases) turns out to be about 0.4, 0.2 and 1.0 hours of the maximum load N_{max}, respectively. The maximum possible storage utilization occurs at the equinox. The following characteristic values for different installed storage capacities in the particular case are apparent:

- possible substituted power of peaking plant N_s during maximum load at mid winter;
- maximum discharge power N_d (for example, storage turbine power) at or near equinox;
- maximum charge power N_c (also at or near equinox).

Actually, the storage utilization will even be better than shown because of the weekends that introduce an added periodicity of the demand.

From an analysis of daily load curves throughout the year (represented by the special cases given in Fig. 1.5) the relative annual stored energy A_{st}/N_{max}, measured in annual hours (h/a) of maximum load, may be found. This is plotted

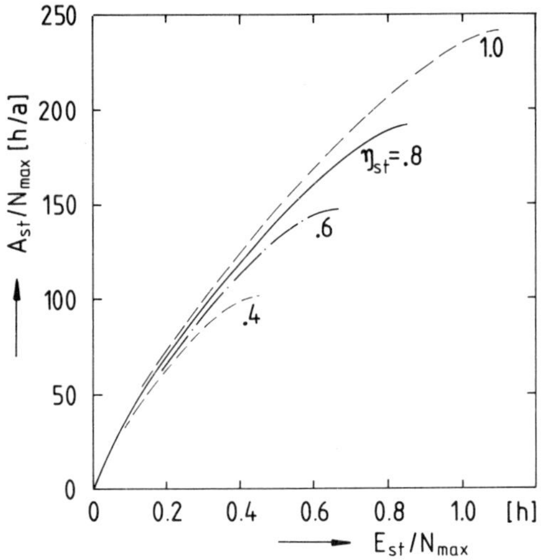

Fig. 1.6. Relative annual stored energy A_{st}/N_{max} versus relative storage capacity E_{st}/N_{max} (parameter: storage efficiency η_{st})

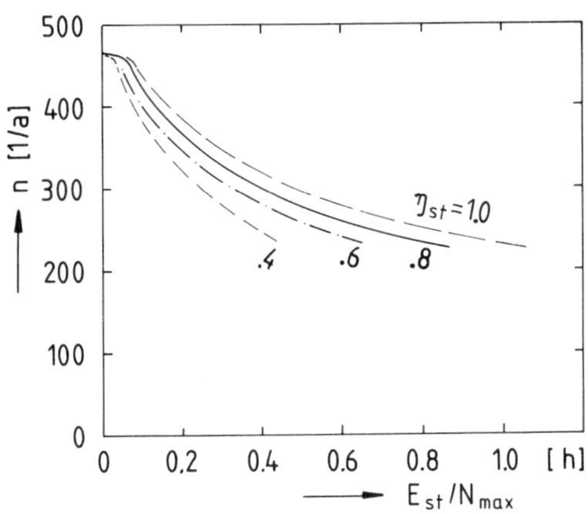

Fig. 1.7. Number of annual discharges n versus relative storage capacity E_{st}/N_{max}

in Fig. 1.6 against E_{st}/N_{max} (h), where E_{st} stands for the total installed storage capacity in the grid. The average number of full discharges per year $n = A_{st}/E_{st}$ of all these storage plants is represented by the slope of the radius vector; and the number of discharges n' of an incremental storage capacity corresponds to the slope of the tangent. A_{st} increases with E_{st} steeply at the beginning, and more slowly at higher values of E_{st} up to a limit at 240 h/a. This, however, is the theoretical upper limit (100% storage efficiency). For storage efficiencies below 1 (for example 0.8, 0.6, 0.4), saturation is reached earlier (at 190, 145, and 105 h/a). This effect of the storage efficiency is the more pronounced the more frequently the use of the storage system is limited by the available charge power.

Fig. 1.7 shows the number of annual discharges n as a function of E_{st}/N_{max} for various storage efficiencies. For a number of storage plants in the grid, E_{st} is the total storage capacity and n is the average number of discharges.

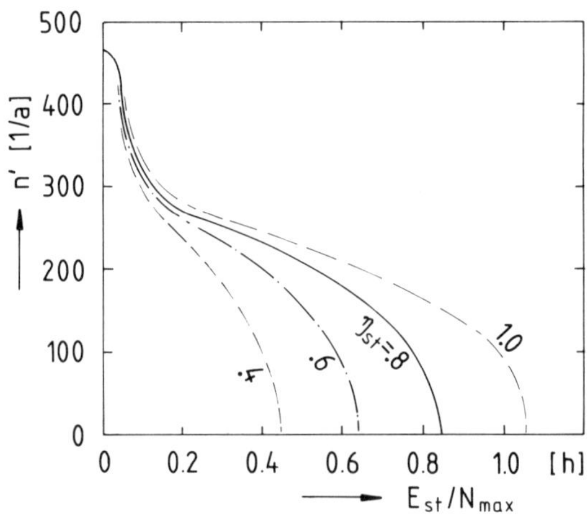

Fig. 1.8. Number of annual discharges n' of a small additional storage plant versus relative storage capacity E_{st}/N_{max}

In Fig. 1.8, the number of discharges n' of a small (incremental) storage plant to be installed in a grid with already existing storage plants of a total capacity E_{st} is plotted according to Fig. 1.7. There is a rather strong dependency on the storage efficiency, the use of the storage systems in the winter season being limited by the available charge energy.

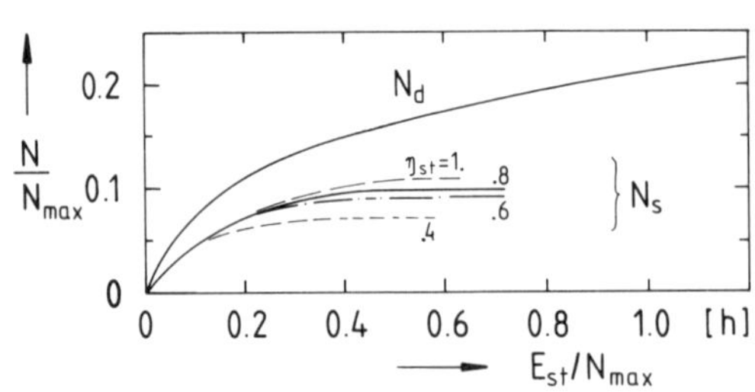

Fig. 1.9. Relative discharge power N_d/N_{max} and relative substituted power of peaking plants N_s/N_{max} versus relative storage capacity E_{st}/N_{max}

Fig. 1.9 shows, against E_{st}/N_{max}, the maximum relative discharge power N_d/N_{max} and also the relative substituted power of peaking plant N_s/N_{max}.

They both can be found from Fig. 1.5 by using the maximum discharge power N_d during the year and the discharge power N_s during the day with the maximum load, respectively. Also N_s depends rather strongly on the storage efficiency.

From N_d one can find the average discharge hours d

$$d = E_{st}/N_d \tag{1.5}$$

and also the incremental discharge hours d'

$$d' = dE_{st}/dN_d. \tag{1.6}$$

Both values are shown below in Fig. 1.10.

Also, from Fig. 1.9, an average ratio

$$R = N_s/N_d \tag{1.7}$$

and an incremental ratio

$$R' = dN_s/dN_d \tag{1.8}$$

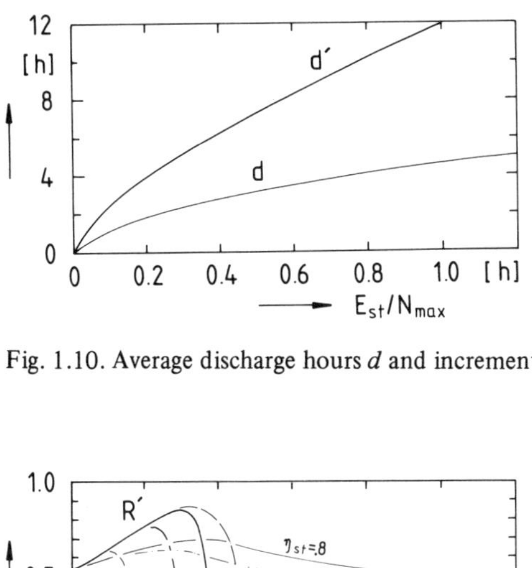

Fig. 1.10. Average discharge hours d and incremental discharge hours d' versus E_{st}/N_{max}

Fig. 1.11. Average peak power substitution ratio $R = N_s/N_d$ and incremental ratio $R' = dN_s/dN_d$ versus E_{st}/N_{max}

may be found which expresses the substituted power of peaking plants in a dimensionless way. In Fig. 1.11 R and R' are plotted against E_{st}/N_{max}.

Now a typical annual cost diagram may be plotted (see Figs. 1.12 and 1.13). Contrary to conventional base load or peaking plants, where investment cost is

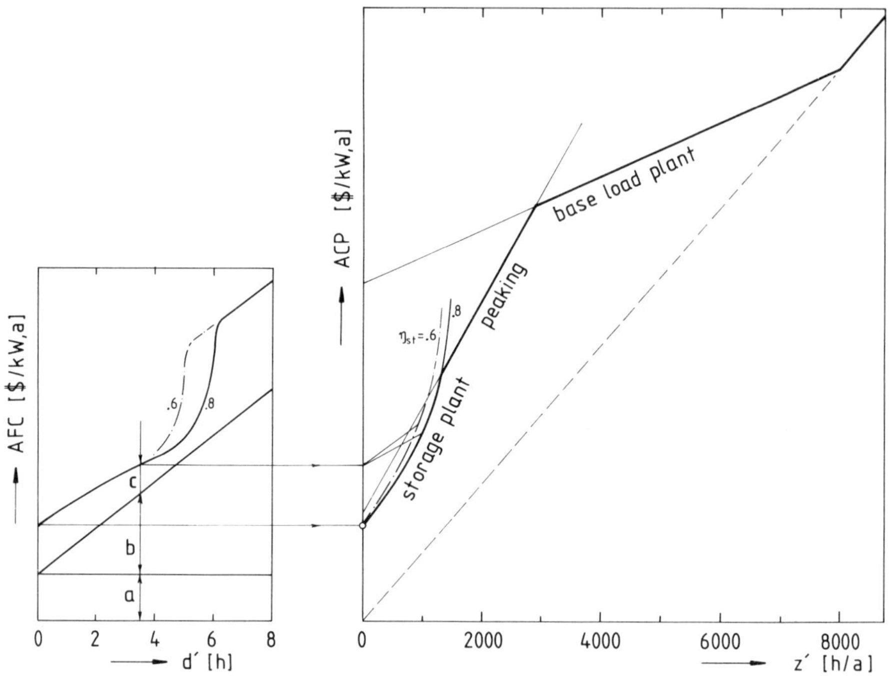

Fig. 1.12. Annual fixed cost AFC of storage plants and nonsubstituted peaking plants versus incremental discharge hours d' of storage plant (left diagram)
a discharge-power-dependent part, *b* capacity-dependent part, *c* cost of nonsubstituted peak power

Fig. 1.13. Annual cost of power generation ACP versus annual load factor z' (right diagram)

only power related and utilization is only limited by technical availability, investment cost of storage plants is power as well as capacity related and utilization is also limited by the demand conditions in the grid.

In Fig. 1.12, which shows the fixed cost of a storage plant, the power-related portion of fixed cost consists of the cost of the peaking turbine, etc.; while the capacity-related portion consists of the cost of storage vessels etc. and the additional cost of the portion $1 - R'$ of the peaking plant that has not been substituted by the storage plant according to Figs. 1.10 and 1.11. This portion, in the particular case shown, decreases slowly at first, then reaches a minimum at $d' = 5$ hours (in the case of $\eta_{st} = 0.8$) and then finally increases again, reaching the installation cost of the peaking plant.

Fig. 1.13 consists of the annual cost lines ($ per kW, per year) versus the annual load factor $z' = n'd'$ (h/a). The base load (for example a nuclear plant) shows high investment cost and low fuel cost; the peaking plant (for instance a gas turbine plant) shows low investment cost and high fuel cost. There may be other plants (for example coal-fired plants) in between.

The line of a storage plant is found by adding the fuel cost of the base load plant to the d'-dependent fixed cost (oblique lines), the fuel cost having been corrected with the relevant efficiencies (the turnaround efficiency of the storage plant, and the part-load efficiency of the base load plant). The actual curve of the storage plant is found by terminating the lines for a given d' at $z' = n'd'$, taking n' from Fig. 1.8 and d' from Fig. 1.10. For simplification, only two storage efficiencies, 80% and 60%, are considered. It turns out that the curve for the storage plant is bent upward. For the case considered, it actually intersects the peaking plant line at about $z' = 1250$ h/a in the situation of 80% storage efficiency.

In this plot, the three negative aspects of low storage efficiency are evident:

- reduction of peak load substitution;
- reduction of use;
- increase of specific fuel cost.

Also, the capacity-dependent first cost increases with decreasing storage efficiency. There is, therefore, a strong incentive for high storage efficiencies.

It should be noted that the straight lines of the fueled plants in Fig. 1.13 are independent of the grid, whereas the line of a storage plant depends on the grid and has to be assessed in each case.

1.3.3. Optimum Portion of Base Load Plants

The procedure discussed in Section 1.3.2 presupposes that the portion of base load plants in the grid is given. This case applies only if, in a given grid with given power plants, an additional storage plant is to be installed. From Fig. 1.13 it also becomes clear in which region of plant utilization a certain type of base load plants and of peak load plants is economic. However, to find the optimum, such an isolated procedure is not sufficient. The feedback of energy storage on the optimum portion of base load plant is to be investigated. For instance, if energy storage is limited by the excess energy of base load plants, then the optimum portion of base load plants may be increased. As before, the question is to which degree the annual energy and peak load will be covered economically by storage plants. However, there is the variable portion of base load plants as an additional parameter.

Fig. 1.14 shows an axonometric diagram of the distribution of annual energy in a typical grid as a function of the base load plant portion N_b/N_{max} and storage capacity E_{st}. The lower plane "A" is the annual energy produced by the base load plants. It is independent of the storage capacity and increases linearly up to the annual minimum load N_{min} (in the example, $N_{min}/N_{max} = 0.25$). For $N_b > N_{min}$, the annual energy is limited by the trapezoidal load, and the slope decreases. Plane "A" intersects plane "C" (the total annual energy) at N_{ave} (in the example, $N_{ave}/N_{max} = 0.50$). To cover the whole annual energy by base load plants plus storage with a storage efficiency of 100%, an installed base load power of N_{ave} was necessary ($\eta_{st} = 1$ has been assumed in the graph).

The area "B", indicating the sum of base load plus storage energy, intersects in the region $N_{min}/N_{max} \leqslant N_b/N_{max} \leqslant 1$. The difference $B - A$ is thus A_{st}; and

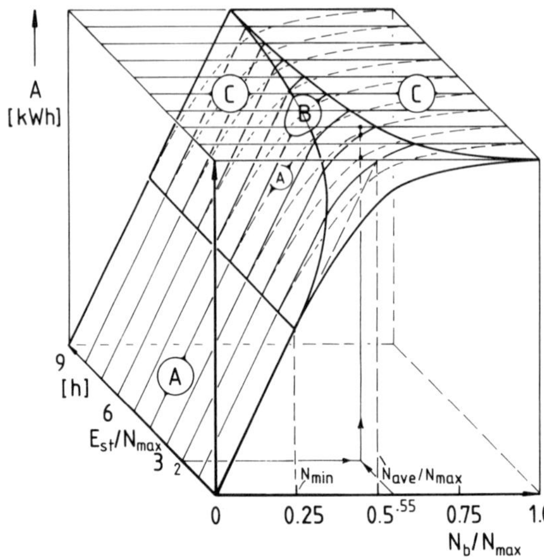

Fig. 1.14. Distribution of annual energy A as a function of relative base load portion $N_b/N_{\max}$ and relative storage capacity $E_{st}/N_{\max}$

the intersection with the given N_b yields a curve that is identical to Fig. 1.6. It increases with increasing storage capacity E_{st} degressively and is limited either by a limitation of excess energy or by a limitation of required discharge. The left-hand limit of area "B" is within the plane "A", the right-hand limit is within plane "C". The following portions of annual energy may be taken from Fig. 1.14, depending on $N_b/N_{\max}$ and E_{st}:

- Base load energy A_b, plane "A".
- Stored energy A_{st}, area "B" minus plane "A".
- Peak load plant energy A_p, plane "C" minus area "B".

There are two bivalent regions and a trivalent one. Their borders may best be found from a vertical projection of Fig. 1.14. This is shown in Fig. 1.15. The following regions are thus apparent:

- Region I ($N_b \leqslant N_{\min}$): $A_{st} \equiv 0$; that is, the load is covered by the base load and the peaking plants only (without storage).
- Region II (the region between "a" and "c"): Here the load is covered by the base load plants, the storage plants, and the peaking plants (the general case). In region IIa, the storage plant is overdesigned; that is, the store is not fully utilized. However, in region IIb (between "b" and "c") this is not the case.
- Region III (between "c" and $N_{\max}$): Here the peaking plants have been completely substituted for by storage plants. The load is covered by the base load plants and storage plants, i.e. by base load plant fuel only. Curve "c" is a limit for $N_b/N_{\max}$; to the right of "c", the installed base load is unnecessarily high (it is assumed that the necessary reserve is accounted for). Case "c" is an energy supply system with one energy converter according to Sec-

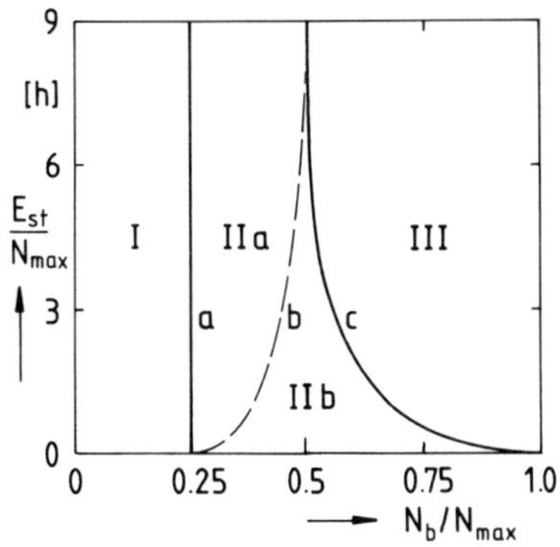

Fig. 1.15. Bivalent and trivalent regions of energy supply as a function of N_b/N_{max} and E_{st}/N_{max}

tion 1.3.1. For very small grids or grids with inexpensive storage, this will be the standard case. It is also the case of complete substitution of peaking plants and their valuable fuel (oil or gas).

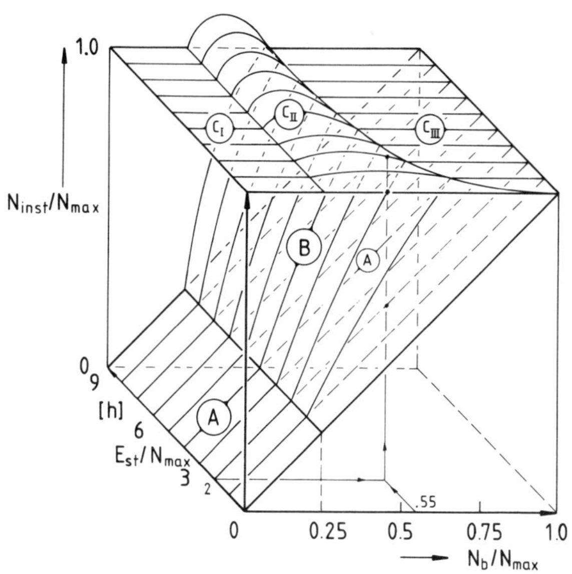

Fig. 1.16. Distribution of installed power N_{inst}/N_{max} as a function of relative base load portion N_b/N_{max} and relative storage capacity E_{st}/N_{max}

Fig. 1.16 shows the distribution of the installed power N_{inst}/N_{max}, as a function of N_b/N_{max} and E_{st}/N_{max}:

- Base load plant power N_b/N_{max} again is a plane "*A*" (now having a slope of 45°).

- The power covered by storage plants, N_d/N_{max}, is the difference between area "B" and plane "A".
- Peaking plant power N_p/N_{max} is the difference between area "C" and area "B". On basis of the regions of Fig. 1.15, C_{II} (the trivalent part of "C") has a hump, since the installed discharge power of the storage plants will not generally be completely available at the time of maximum demand N_{max}, as has been explained in Section 1.3.2 and in Figs. 1.9 and 1.11.

The total annual cost of power generation (TACP) may now be assessed from the distribution of annual energy in Fig. 1.14 and of installed power in Fig. 1.16:

$$\mathrm{TACP} = A_b b_b + A_{st}(b_b/\eta_{st}) + A_p b_p + N_b c_b a_b + N_d c_{st} a_{st} + N_p c_p a_p \tag{1.9}$$

where

b	fuel cost
η_{st}	storage efficiency
c	specific capital cost
a	annuity.

Indices:

b	base load
st	storage
p	peak load
d	discharge.

The value of TACP (the cost minimum being marked 100%) is plotted in a shell diagram (Fig. 1.17) for assumed values of capital and fuel cost, in its dependence on the base load plant portion N_b/N_{max} and the relative storage capacity E_{st}/N_{max}. The following results are found.

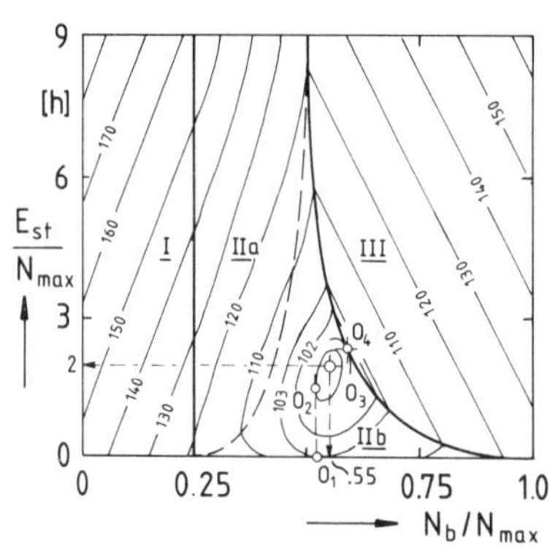

Fig. 1.17. Total annual cost of power generation TACP as a function of relative base load portion N_b/N_{max} and relative storage capacity E_{st}/N_{max}
O_3 absolute minimum of TACP (= 100%)

- In region I (base load plant plus peaking plant only), the cost decreases with increasing N_b/N_{max}. It increases with E_{st}/N_{max}, since no storage plant can operate, because there is no surplus energy from base load plants available.
- In region II, the relative optimum (cost minimum) O_1 for $E_{st} = 0$ (that is, between the base load plant and the peaking plant power) is found at $N_b/N_{max} = 0.52$ ($N_p/N_{max} = 0.48$). If this value $N_b/N_{max} = 0.52$ is kept constant and E_{st} is varied, the relative optimum O_2 is found at $E_{st} = 1.5$. The absolute optimum O_3 is at $N_b/N_{max} = 0.55$ and $E_{st}/N_{max} = 2.0$ (and, according to Fig. 1.16, $N_d/N_{max} = 0.33$ and $N_p/N_{max} = 0.14$).
- There is another relative optimum O_4 in region III (base load plant and storage only, $N_p/N_{max} = 0$) on the curve "C", at $N_b/N_{max} = 0.58$ and $E_{st}/N_{max} = 2.4$.

Therefore, the optimum size of storage plants is, in general, an optimization problem with two parameters. The daily load shapes for the entire year (or for several years), the storage efficiency, and the storage operation are the input, the optimum values of the storage capacity E_{st}, of the relative base load plant power N_b/N_{max}, of the storage discharge power N_d/N_{max}, and of the peak load plant power N_p/N_{max} are the output.

1.4. The History of Thermal Energy Storage (TES)

Although thermal energy storage (TES) was applied to building design and to heating systems at a very early date, industrial applications began only in the 19th Century: Regenerators for flue gas heat recovery by means of ceramic materials have been used since about the middle of the 19th Century in industrial processes requiring high combustion temperatures and, hence, high air temperatures. This system is still in use in the metallurgical and glass industries, in cokeries, in the low temperature part of steam generators, and, more recently, in open-cycle gas turbines as well as in air-conditioning systems.

Applications of TES to the storage of mechanical energy for vehicles by means of thermochemical (NaOH) storage date back to 1880: A river boat was built in 1883 and sailed on the Spree; a street car was in operation between Berlin and Charlottenburg; and the railway locomotive of Honigmann did regular duty, hauling passenger trains between Jülich and Aachen in 1882 [8.6]. This locomotive had no exhaust. Waste steam was absorbed by the NaOH, generating heat to evaporate steam in boiler tubes (Fig. 1.18).

A charging station reconcentrating the NaOH by evaporation was located at one of the railroad stations (Fig. 1.19).

The NaOH storage process has since been forgotten and only recently was proposed again [8.18].

Several years before the Honigmann locomotive, the principle of steam storage (in pressurized saturated water) was applied to street cars in New Orleans. Storage pressure was equal to the usual operating pressure. The power decreased with the sliding storage pressure. In industrial plants, where a furnace would not be permitted for safety reasons, locomotives of this type were frequently used

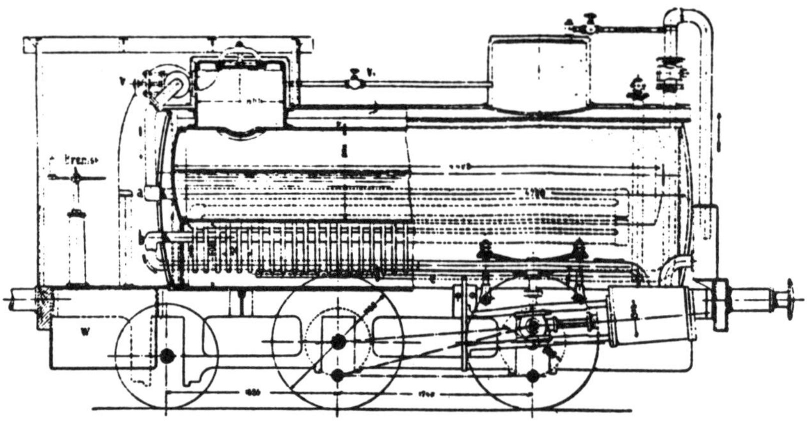

Fig. 1.18. Thermochemical locomotive of Honigmann [8.7]

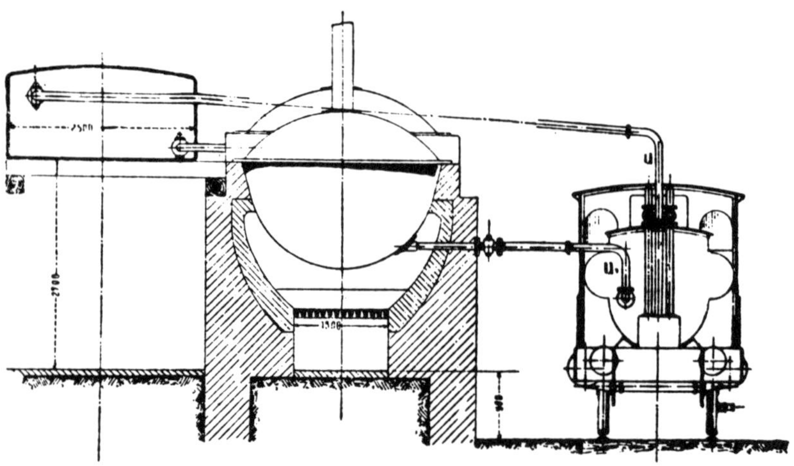

Fig. 1.19. Charging station of the Honigmann locomotive [8.7]

and became known as "fireless locomotives" (see Fig. 1.20).

Applications of TES to stationary steam generation date back to about the same time: The external hot water vessel for steam storage (see Fig. 1.21)–a precursor to the sliding-pressure storage of Ruths at the beginning of the 20th Century–was patented by McMahon in 1873 [1.20].

The development of the present-day sliding-pressure steam storage scheme began with the work of J. Ruths. A German patent, granted to Ruths in 1913 [1.20], had already indicated applications to power generation. The first steam storage power plant with peaking turbine was built in 1920 at Malmö, Sweden, for peaking and reserve duty. Also, the feed water storage system (displacement storage at constant pressure) was developed in the 1920s by Marguerre, and a

Fig. 1.20. "Fireless" locomotive [8.10]

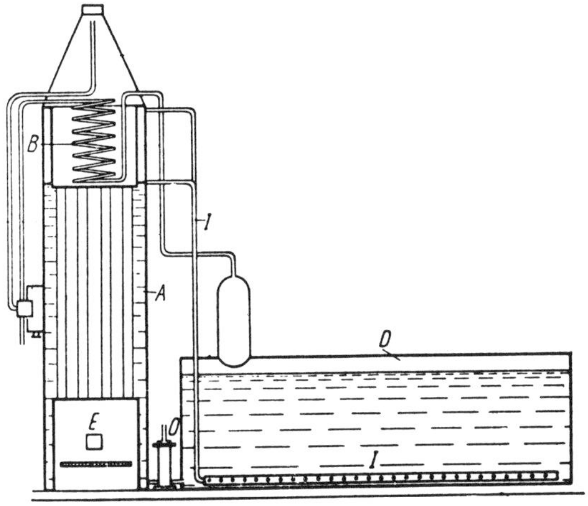

Fig. 1.21. Storage boiler invented by MacMahon [1.20]
A boiler, *B* superheater, *D* accumulator (store), *E* furnace, *I* charging pipe, *O* pump

large plant was built at Mannheim. The largest steam storage power plant was built in 1929 in Berlin-Charlottenburg, with an electric power of the peaking turbine set of 50 MW, a storage capacity of 67 MWh and a storage pressure of 14 bar (200 psi). After more than 50 years, this sliding-pressure storage plant is still on day-and-night-peaking-and-reserve duty in the Berlin island grid. Fig. 1.22 exhibits one of the 16 riveted pressure vessels, while Fig. 1.23 shows the general arrangement.

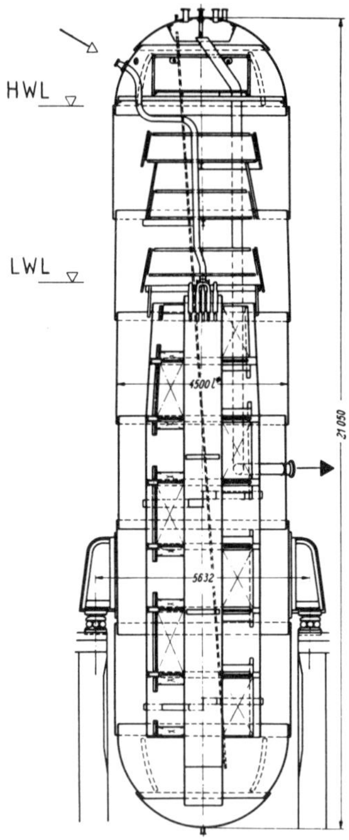

Fig. 1.22. One of the 16 riveted steel vessels of the Berlin-Charlottenburg plant [1.20]–dimensions in millimeters
HWL water level at 14 bar, fully charged, *LWL* water level at 1.5 bar, fully discharged

In the 1920s, sliding-pressure storage was developed for high pressures (up to 120 bar) and for superheated steam by P. R. Gilli. The high pressure steam generated in the storage vessel is throttled to working pressure (14–17 bar) and is subsequently superheated in the steam cushion of the storage vessel [2.4, 8.8–8.15]. This system was applied to a large number of locomotives (Fig. 1.24) and to the instantaneous reserve plant at the Simmering steam power plant of the Vienna Municipality (Fig. 1.25).

Industrial TES plants were built in rather large numbers, mostly according to the sliding-pressure principle. In particular, they were used in the chemical, food, and metallurgical industries for process heat applications. Displacement storage systems were also built for district heating applications.

Renewed interest in TES stems mainly from the need to save and substitute oil products in heat and power generation. Solar facilities for space heating, hot tap water production or process heat generation will need TES in almost all cases, district heating and heat pump installations in many cases. Solar power

Fig. 1.23. The Berlin-Charlottenburg peak load storage plant

Fig. 1.24. First high-pressure steam storage locomotive (the Gilli locomotive) of 1934, 118 bar, 82 tonnes

Fig. 1.25. Instantaneous reserve plant at Simmering steam power plant, Vienna, Austria

plants do require energy storage for reserve and bottoming applications. For base load plants, such as coal or nuclear plants, substituting oil-fired peak load plants by means of energy storage will mean operational advantages and better plant utilization. In particular, nuclear plants (and even more so, future breeder and fusion plants) with their high capital, low fuel costs are candidates for integrated or pumped energy storage whenever the total power of high capital, low fuel cost plants (nuclear, river hydro, solar etc.) exceeds base load.

Such new applications as well as new vessel technologies (for example, underground caverns and Prestressed Cast Iron Vessels) do require new storage principles and flow sheets. The Compressed Air Energy Storage (CAES) system, an open-cycle gas turbine with a pneumatic pumped storage system using underground storage, was developed and built [7.1–7.11]. For steam power plants, the integrated TES system using expansion-type accumulators and a combined steam and feed water storage was developed [4.23].

2. Basics

Some definitions:

- *Thermal energy storage* (TES) is a physical or chemical process taking place in the store (accumulator) during the charge and discharge operation.
- The *store* consists of the storage vessel (usually thermally insulated), the storage medium, the charging and discharging devices, and the auxiliaries.
- The *storage system* is defined by the manner in which energy for charging the accumulator is extracted from the energy source and in which energy is discharged from the accumulator and is, in many cases, transformed into the required form of energy.

2.1. Classifications

Fig. 2.1 shows a TES process, the boundaries of which are given by the storage vessel. A formal energy balance may be written as

$$in - ex = st \tag{2.1}$$

where "*in*" means energy input, "*ex*" energy output, and "*st*" storage [2.1, 2.2].

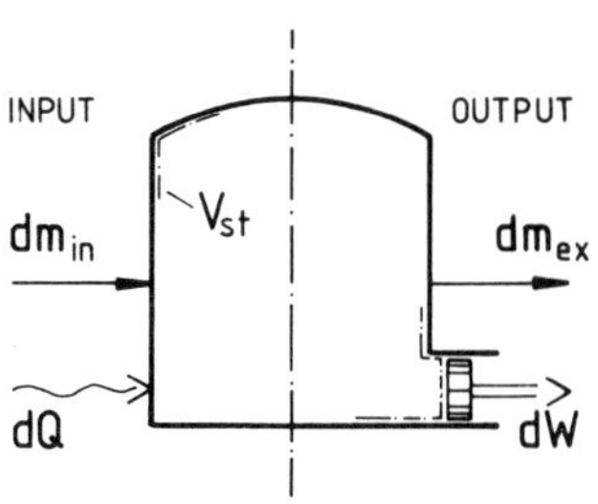

Fig. 2.1. Energy balance of an accumulator

If the First Law of Thermodynamics is applied to the energy input and output of this open system, the "general equation of energy storage" for open systems is then obtained

$$(u + pv + gH + c^2/2)_{in}\, dm_{in} + dQ - (u + pv + gH + c^2/2)_{ex}\, dm_{ex} - dW$$
$$= d[(u + gH + c^2/2)_{st} m_{st}] \qquad (2.2)$$

where

m_{st}	mass of storage medium in storage vessel
dm_{st}	differential change of m_{st}
dm_{in}	differential mass input
dm_{ex}	differential mass output
u	specific internal energy (counted from an arbitrary datum level)
p	pressure
v	specific volume
pv	specific energy necessary for input into or output from the system of unit mass; "flow energy"
g	acceleration of gravity
H	height (related to an arbitrary reference level)
gH	specific potential energy
c	velocity
$c^2/2$	specific kinetic energy
dQ	heat transmitted independently of mass transfer (because of Eq. (2.2), $+dQ$ is the heat added to the system)
dW	work of the system, independent of mass transfer; for example, movement of system borders, electrical energy, shaft power of an engine (because of Eq. (2.2), $+dW$ is the work done by the system).

Inspection of the general equation of energy storage reveals that the energy storage capability may be due to a change of

(1) specific internal energy, or
(2) specific potential energy, or
(3) specific kinetic energy, or
(4) mass of the system.

Usually, case (1) is called thermal energy storage (in the strict sense), and also case (4), provided that the specific internal energy is higher than that of the ambient.

In this book, the definition of thermal energy storage is, however, extended to include case (4) even if the specific internal energy is low but the exergy (available energy) is high because of the state of the storage medium. This applies, for instance, to the pressurized gas (air) storage systems that are capable of storing a substantial amount of exergy even if the gas has ambient temperature and thus negligible internal energy. It is evident that such a thermal storage system is utilized in power storage only, whereas thermal storage in the strict sense may be used for power as well as for heat applications.

If both the kinetic energy storage and the potential energy storage are excluded ($c_{st} = 0$; $H = 0$), and if, furthermore, the kinetic and potential energy term of mass input/output is neglected and the work is restricted to movement of the system's boundaries, that is if

$$dW = p_{st}\, dV_{st} \tag{2.3}$$

with

V_{st} storage volume and
p_{st} storage pressure,

then Eq. (2.2) is reduced to a form which is valid for thermal energy storage

$$(u + pv)_{in}\, dm_{in} + dQ - (u + pv)_{ex}\, dm_{ex} = d(um)_{st} + p_{st}\, dV_{st}. \tag{2.4}$$

Using the definition of enthalpy, we have

$$h = u + pv \tag{2.5}$$

and the thermal energy balance thus becomes

$$h_{in}\, dm_{in} + dQ - h_{ex}\, dm_{ex} = d(um)_{st} + p_{st}\, dV_{st}. \tag{2.6}$$

The mass balance is

$$dm_{in} - dm_{ex} = dm_{st}. \tag{2.7}$$

The charging and discharging process is described in general terms by Eqs. (2.4) or (2.6), and (2.7). In the simple cases, analytic solutions may be possible. Otherwise, numerical evaluations may be performed, this being particularly true for the discharge operation.

According to the above definitions and deductions, thermal energy storage may be classified according to the following categories:

(a) *Storage Medium and Heat Transfer Medium.*

(a1) Direct Storage: Storage medium and heat transfer medium are identical. The storage medium may be solid (a moving bed), liquid, in the two-phase (liquid plus vapor) state, or gaseous.

(a2) Indirect Storage: Thermal energy is transformed either by heat transfer only (for instance, by heat conduction through the vessel wall) or by means of mass transfer by a separate heat transfer medium (in the liquid, two-phase, or gaseous state). The storage medium itself may be solid, liquid or gaseous without phase change, or may employ a phase change, either solid/solid, or solid/liquid, or liquid/vapor.

(a3) Semi-direct Storage: As in (a2) except that the storage capacity of the heat transfer medium plays a more important role (for instance, hot oil storage with a solid bed).

(a4) Sorption Storage: The ability of some storage media to absorb gases and thereby release heat (and to absorb heat when they release the gas) is used here. The heat transfer may be direct or via the gas.

(b) *Mass of Storage Medium.*

(b1) Constant Mass Content ($dm_{st} = 0$): This is usually a case of indirect storage. However, there may be direct storage if the released mass after cooling down (during discharge) or heating up (during charging) is completely returned to the store (displacement storage).

(b2) Variable Mass Content ($dm_{st} \neq 0$): This is always a case of direct storage.

(c) *Storage Volume.*

(c1) Constant Volume ($dV_{st} = 0$): This applies to storage in closed vessels, at least as an approximation.

(c2) Variable Volume ($dV_{st} \neq 0$): This situation may apply to storage at ambient pressure or with special pressurization equipment.

(d) *Storage Pressure.*

(d1) Constant Pressure ($dp_{st} = 0$).

(d2) Varying (Sliding) Pressure ($dp_{st} \neq 0$).

2.2. Storage of Sensible Heat in Saturated Fluids

Storage arrangements of this kind consist of a liquid in the lower part and vapor in the upper part, the liquid and vapor being in thermodynamic equilibrium (at saturation temperature). The storage medium is almost exclusively water/steam. The storage vessel may contain a small amount of liquid or it may be almost completely filled with liquid. Fig. 2.2 shows an arrangement with all possible kinds of discharge and charge equipment. There may be discharge of saturated steam via the discharge line (4), of superheated steam via a throttling valve and the discharge line (5), of liquid via the discharge line for hot water (7), or of heat only by means of the heat exchanger surface (6). In this latter case the storage mass remains constant (indirect sliding-pressure storage).

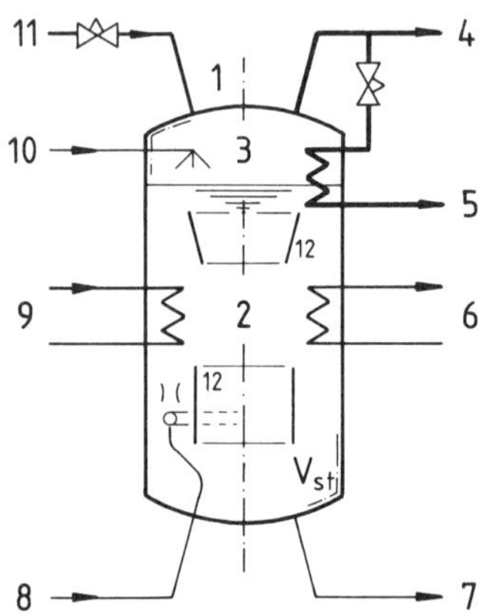

Fig. 2.2. Sliding-pressure accumulator
1 pressure vessel, *2* water content, *3* steam cushion, *4* discharge line for saturated steam, *5* discharge line for superheated steam, *6* discharge heat exchanger surface, *7* discharge line for hot water, *8* charge line for steam, *9* charge heat exchanger surface, *10* charge line for (hot) water, *11* charge line for heating steam, *12* internals

Charging can be carried out by admission of (hot) water via the charge line (10) which may be heated up to saturation by condensing the steam that comes through the charge line (11), by adding heat to the system by means of the heat exchanger surface (9), or by blowing steam through line (8) in the water content. Internals (12) will minimize stratification.

Change of pressure during charging and discharging is characteristic of this type of storage. Change of phase (condensation, evaporation) will always take place, but it does not have much effect on storage capacity which is determined largely by the change of sensible heat. Therefore, saturated fluid storage is regarded as a special type of sensible heat storage.

2.2.1. Sliding-pressure Storage

For sliding-pressure storage (varying-pressure storage, steam storage) with discharge of saturated steam, only the discharge line (4) in Fig. 2.2 is used. The storage vessel is a pressure vessel and is usually thermally insulated. About 75 to 95% of the vessel volume is taken up by saturated water, while 5 to 25% is taken up by the steam "cushion".

The discharge operation is initiated by opening a valve in the discharge line. Steam is produced in the liquid by a continuous decrease of pressure. The steam cushion serves as a means of phase separation. Saturated or wet steam with only little liquid content (typically less than 1% by weight) leaves the vessel. Because of the high enthalpy of the steam, only a relatively small portion of the liquid is evaporated during discharge, the main part remaining in the vessel at the lower temperature.

To avoid temperature stratification during charging or discharging and to achieve the necessary circulation of the water volume, suitable internals (12) are provided, this being particularly true for vertical vessels (see Section 4.2).

Charging is usually achieved by blowing saturated or superheated steam of a somewhat higher pressure into the liquid in such a way that circulation is generated by buoyancy and ejector effects [2.3]. Charging is also possible via a heat exchange surface with steam of a temperature higher than that of the liquid; that is to say, by condensing saturated steam of higher pressure or by cooling superheated steam of a sufficient temperature (even of lower pressure: "steam transformer").

The discharging of this type of storage vessel always means a considerable decrease of pressure (and charging means a corresponding increase). Therefore, this type of storage is called "sliding-pressure storage" (as opposed to the "constant-pressure storage" system). Sometimes, the name "Ruths" is given to it, this name being for the inventor J. Ruths who introduced this system during the 1920s to various branches of industry as well as to steam power plants.

The advantages of this system include: immediate availability, relatively low investment cost, direct supply of steam, and charging with steam. New applications of this type of system seem possible, given the new pressure vessel technologies and the correspondingly new flow sheets. Thermal storage systems of this kind have also been proposed for large underground storage caverns (see Section 4.4).

Discharge is described by Eq. (2.6), where $dm_{in} = 0$ and, as an approximation, $dV_{st} = 0$. If the steam is assumed to be saturated, then Eq. (2.6) becomes

$$- h''_m \; \Delta m_{ex} + \Delta Q = \Delta(u_{st} m_{st}) \tag{2.8}$$

where Δ stands for finite differences, for instance

$$\Delta(u_{st} m_{st}) = (u_{st} m_{st})_e - (u_{st} m_{st})_b \tag{2.9}$$

b beginning of the process (discharge)
e end of the process
$''$ saturated steam.

The index m in Eq. (2.8) refers to the mean value of enthalpy. For sufficiently small increments, $h''_m = (h''_b + h''_e)/2$.

The mass balance (2.7) reduces to

$$- \Delta m_{ex} = \Delta m_{st} = m_{st,e} - m_{st,b} \tag{2.10}$$

Eq. (2.8) cannot be evaluated in a straightforward manner if the influence of the steam volume in the vessel cannot be neglected. In such cases (for instance, at high pressures), not only the water content at u' is to be considered but also the steam content at u''

$$u_{st} = u'(m'/m_{st}) + u''(m''/m_{st}) \tag{2.11}$$

$'$ saturated water.

From a volume balance

$$V_{st}/m_{st} = v'm'/m_{st} + v''m''/m_{st} \tag{2.12}$$

and from a mass balance

$$m_{st} = m' + m'' \tag{2.13}$$

follows the water content

$$m'/m_{st} = 1 - m''/m_{st} \tag{2.14}$$

and the steam content

$$m''/m_{st} = (V_{st}/m_{st} - v')/(v'' - v') \tag{2.15}$$

and Eq. (2.11) thus becomes

$$u_{st} = u' + (V_{st}/m_{st} - v')(u'' - u')/(v'' - v'). \tag{2.16}$$

The right-hand side consists of a part with specific intensive variables of state only and with an extensive part. With

$$j = u' - v'(u'' - u')/(v'' - v') \tag{2.17}$$

Eq. (2.16) becomes

$$u_{st} = j_{st} + (V_{st}/m_{st})(u'' - u')/(v'' - v') \tag{2.18}$$

and the storage equation (2.8) becomes

$$h'' \; \Delta m_{ex} = (j_{st} m_{st})_b - (j_{st} m_{st})_e + V_{st} \left[\left(\frac{u'' - u'}{v'' - v'} \right)_b - \left(\frac{u'' - u'}{v'' - v'} \right)_e \right] + \Delta Q \tag{2.19}$$

or with the mass balance as given in Eq. (2.10), we have

$$
\begin{aligned}
(h''_m - j_{st,e})\,\Delta m_{ex} &= \\
&= (j_{st,b} - j_{st,e})m_{st,b} + V_{st}\left[\left(\frac{u''-u'}{v''-v'}\right)_b - \left(\frac{u''-u'}{v''-v'}\right)_e\right] + \Delta Q \\
&= \quad a \quad + \quad b \quad + c. \qquad (2.20)
\end{aligned}
$$

It is common practice—even if it is a practice that is only useful for flowing media—to use the enthalpy h rather than the internal energy u. With Eq. (2.5) and

$$r = h'' - h' \tag{2.21}$$

$$w = v'' - v' \tag{2.22}$$

we obtain

$$\frac{u''-u'}{v''-v'} = \frac{(h''-h') - p(v''-v')}{v''-v'} = \frac{r}{w} - p \tag{2.23}$$

and

$$j_{st} = [u' - v'(u''-u')/w]_{st} = (h' - v'r/w)_{st}. \tag{2.24}$$

Eq. (2.20) may therefore be written as

$$
\begin{aligned}
(h''_m - j_{st,e})\,\Delta m_{ex} &= \\
&= (j_{st,b} - j_{st,e})m_{st,b} + V_{st}[(r/w)_b - (r/w)_e - (p_b - p_e)] + \Delta Q \\
&= \quad a \quad + \quad b \quad + c. \qquad (2.25)
\end{aligned}
$$

The storage equation according to Eqs. (2.20) or (2.25) thus accounts for:

- the change of energy content of the storage vessel including its steam volume by using j instead of u according to term (a) and the additional term (b);
- the heat supplied (for example, by thermal energy storage in the vessel wall and/or via an additional heat exchange surface) minus heat losses by the term (c).

This equation has been deduced by P. R. Gilli in 1933 [2.4] and it was extended by G. Beckmann [2.5, 2.6] by a term for vessel dilatation. Hardegen [2.7] has considered the effect of not fully degassed water.

In evaluating the storage equation for a change of state from "b" to "e" (for instance from p_b to p_e), small increments of p are assumed, and the mass output Δm_{ex} is assessed. For the next step, $m_{st,b}$ is reduced by Δm_{ex}.

Fig. 2.3 shows the specific mass storage capacity in a $\rho(t)$ diagram. The density ρ_{st} of the water/steam content is plotted against saturation pressure p_s or saturation temperature t_s, with the relative steam volume y as parameter.

The state at the beginning of the discharge ("b") is usually defined by p_b and y_b. Thus $\rho_{st,b}$ is immediately determined. The discharge is represented by the curves down to the final pressure. Point "e" defines density and steam volume at the end of the discharge. The steam output per unit vessel volume, the "steam rate", is given as $\Delta\rho_{st} = \rho_{st,b} - \rho_{st,e}$.

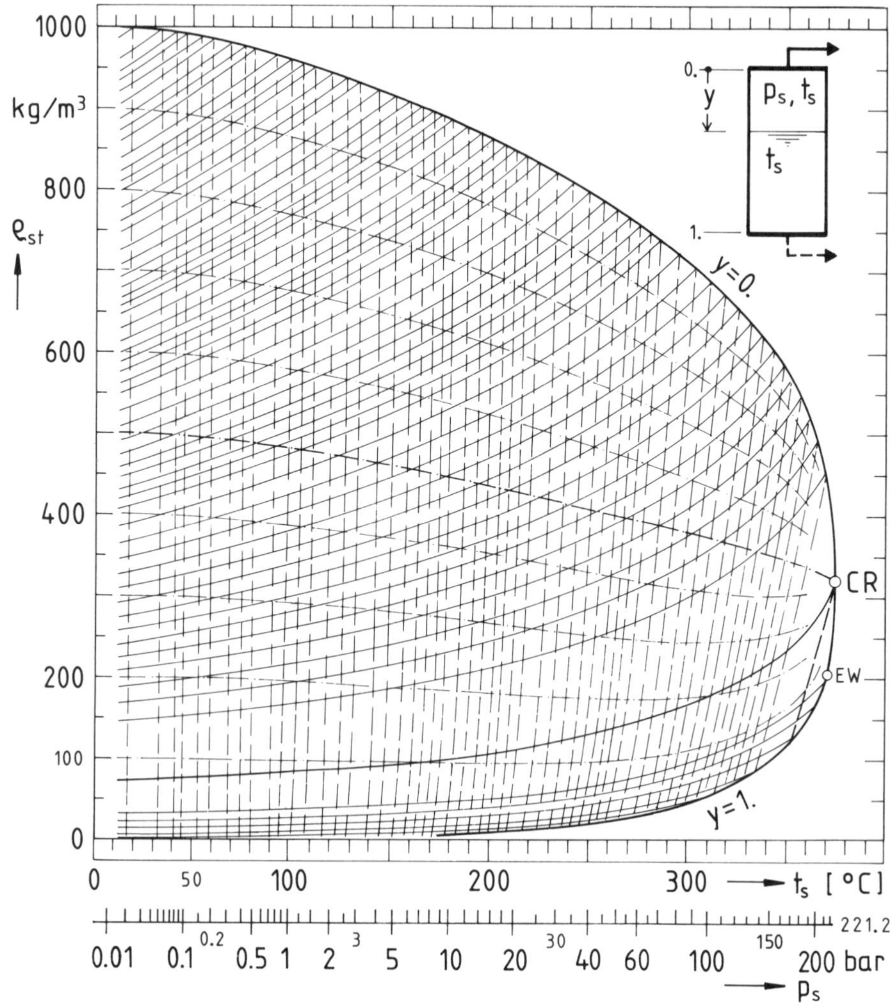

Fig. 2.3. Specific mass storage capacity of a sliding-pressure accumulator (storage density ρ_{st} versus saturating temperature t_s or p_s)
Unbroken line: discharge of saturated steam, Dashed line: discharge of saturated water, *CR* critical point, *EW* end of water discharge, *y* steam volume
Example for steam discharge:
Begin of discharge: p_b = 60 bar (t_b = 275°C), y_b = 0.10
End of discharge: p_e = 10 bar (t_e = 180°)
Results: $\rho_{st,b}$ = 685 kg/m^3, $\rho_{st,e}$ = 535 kg/m^3, y_e = 0.60
Steam rate: $\Delta\rho_{st}$ = 685 − 535 = 150 kg steam per m^3 storage volume

In this diagram, the upper limiting curve represents the density of water at saturation (ρ') and the lower one the density of saturated steam (ρ''). The limits of increasing the storage capacity (at given discharge pressure) by very high charge pressure thus become evident: The enthalpy increases continuously, but the density decreases considerably.

Fig. 2.3 is based on Eq. (2.25) and on the Steam Tables [2.8]. No heat to or from the inner vessel surface—term c in Eq. (2.25)—was considered. However, the capacity of high pressure storage vessels is markedly increased by heat stored in the vessel wall [1.18]. As an approximation, this effect may be accounted for by counting the volume of the vessel wall as water volume, since the volumetric heat capacities of steel and water are approximately equal.

Fig. 2.4 shows the specific heat storage capacity of the sliding-pressure storage system. The upper curve has a maximum at approximately 127 bar. Higher pressures thus tend to be uneconomic for mere heat storage applications.

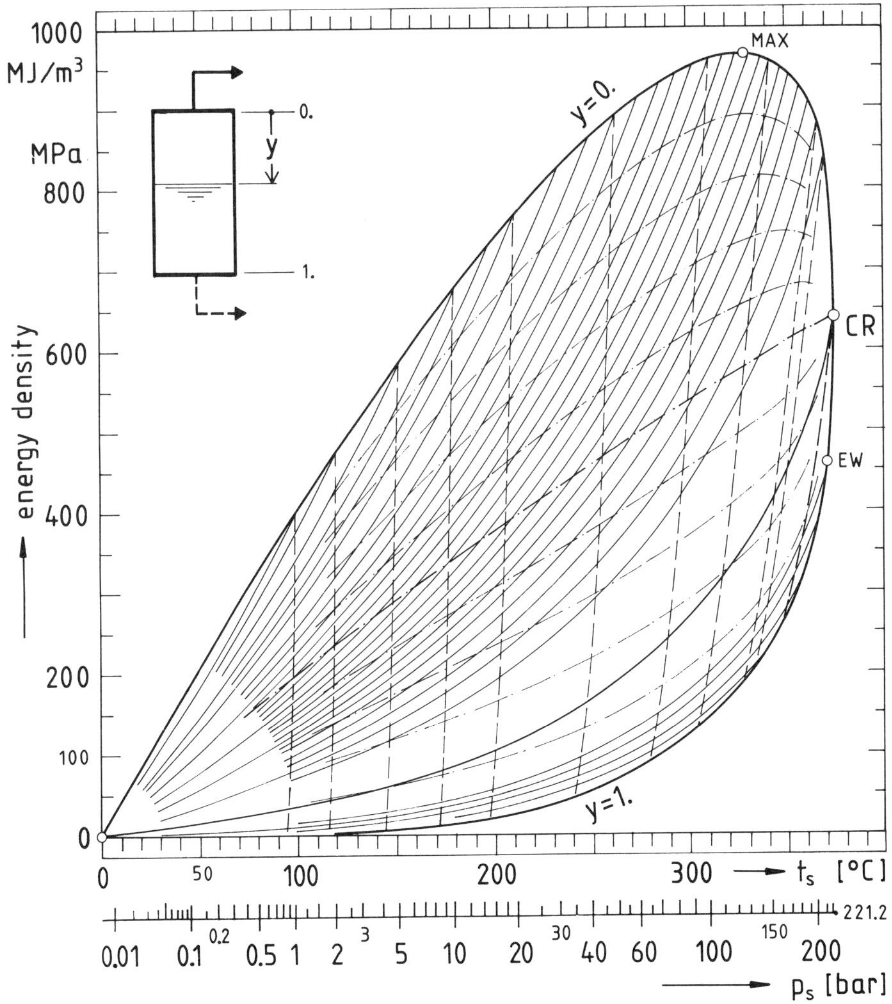

Fig. 2.4. Specific heat storage capacity of a sliding-pressure accumulator (energy density versus t_s or p_s)—legend as in Fig. 2.3

Example as per Fig. 2.3: Energy density = 830 – 410 = 420 MJ/m^3 or MPa

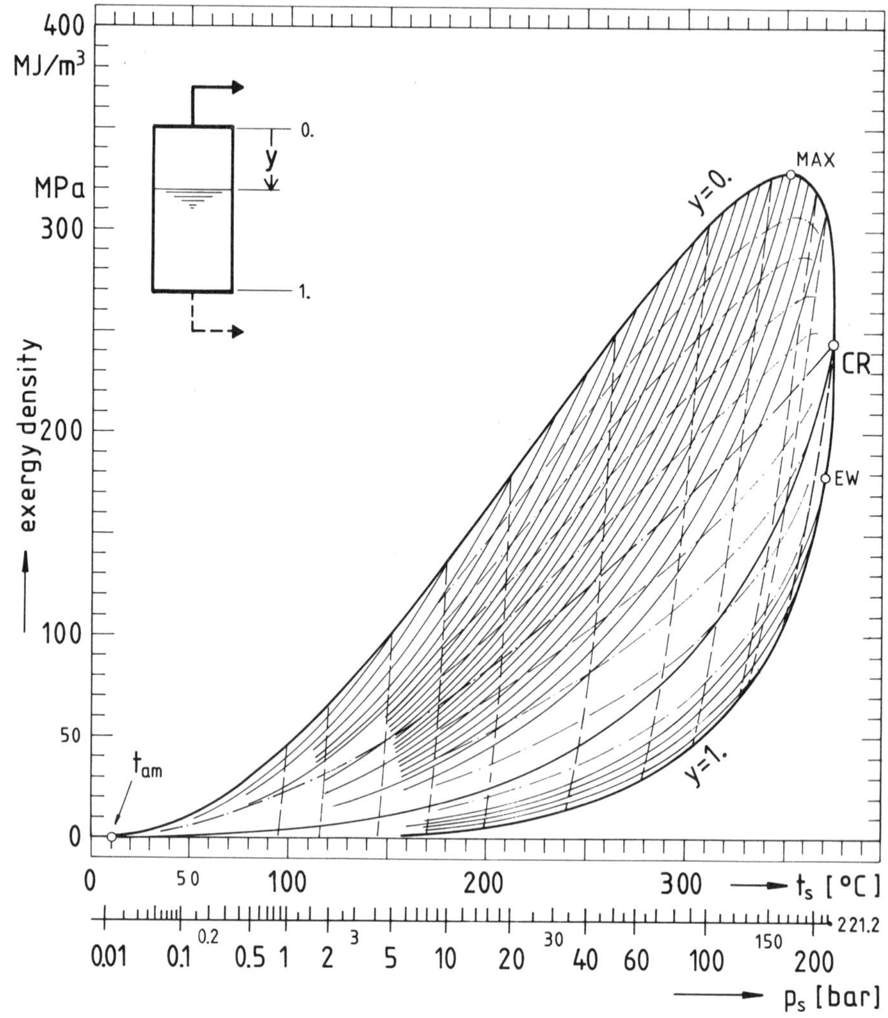

Fig. 2.5. Specific exergy storage capacity of a sliding-pressure accumulator (exergy density versus t_s or p_s)–legend as in Fig. 2.3
Example as per Fig. 2.3: Exergy density = 240 − 85 = 155 MJ/m³ or MPa

However, if heat is stored for power plant applications, it is neither the mass storage nor the heat storage but the storage of available energy (the exergy) that counts. Fig. 2.5 shows the specific exergy storage capacity (the exergy density) for an ambient temperature t_{am} = 10°C (283 K), corresponding to a saturation pressure of 0.012 bar. Indeed, it turns out that high pressures show some incentive for power applications (the maximum is only reached at approximately 170 bar or 352°C).

If a sliding-pressure accumulator has a discharge for superheated steam (see line (5) in Fig. 2.2), Eqs. (2.20) and (2.25) apply to this case if the appropriate enthalpy of the superheated steam is taken for h''_m.

2.2.2. Steam Cushion Storage

This type of steam storage is presently being used for catapults (for example, in air-craft carriers). Similar to the sliding-pressure storage system, there is a thermally insulated pressure vessel which is completely filled with steam ($y = 1.0$) at the beginning of the discharge. If the steam is saturated, then part of it condenses during discharge because of the expansion, thus leading to some water volume in the vessel. The discharge rate may be very high.

This type is a special case of the sliding-pressure storage system, having $y_b = 1$. Discharge lines starting from $y = 1$ have therefore been introduced into Figs. 2.3, 2.4 and 2.5 (the lines below the critical point CR). Specific storage capacity is much lower than that of the sliding-pressure type with water content. Storage capacity of the vessel wall is thus of a relatively high significance. However, heat transfer from vessel to steam is moderate during discharge, and thermal equilibrium between wall and steam cannot, therefore, be assumed. In particular, this is true if discharge is rapid.

Storage of pressurized gases and highly superheated steam will be dealt with in Section 2.6.

2.2.3. Expansion-type Storage

This type is similar to the sliding-pressure storage system in that the pressure vessel is filled with saturated water. The steam volume may even be smaller. However, the discharge line is situated in the lower part of the vessel. If the discharge valve is opened, then pressurized hot water is discharged through the line (7) in Fig. 2.2. The steam volume in the vessel increases. The additional steam is produced by evaporation of a small part of the water volume, thus causing the pressure to decrease slightly.

The charging of the system is achieved either by applying pressurized hot water which in the earlier part of the charging operation generates steam which is compressed and later partly condensed. Despite some mixing (exergy) losses, it may be preferable to charge at constant pressure simultaneously with subcooled hot water through line (10) and with steam through line (11).

Expansion-type storage systems have so far been used for rockets (see Chapter 8), for ejectors in intermittent supersonic channels, and for pressurizers for closed hot water systems. Moreover, applications for superheating stored steam and for very large pressure vessels (because of low temperature gradients during charging and discharging) have been proposed [2.9].

Eq. (2.20), which is valid for sliding-pressure storage, is used for an assessment of storage capacity. However, the enthalpy of the discharged mass is now that of saturated water h'_m rather than saturated steam h''_m

$$(h'_m - j_{st,e})\,\Delta m_{ex} = (j_{st,b} - j_{st,e}) m_{st,b} + V_{st}\left[\left(\frac{u'' - u'}{v'' - v'}\right)_b - \left(\frac{u'' - u'}{v'' - v'}\right)_e\right] + \Delta Q. \tag{2.26}$$

Storage capacities of the expansion-type storage are shown in Figs. 2.3, 2.4 and 2.5 by dotted lines.

The mass output per unit vessel volume (Fig. 2.3) is considerably higher than in the case of sliding-pressure storage; the energy density (Fig. 2.4) and exergy density (Fig. 2.5) are similar; and the pressure and temperature swings are much smaller. Following the water discharge at nearly constant pressure, a discharge of steam plus some condensate (down to some low pressure) may follow.

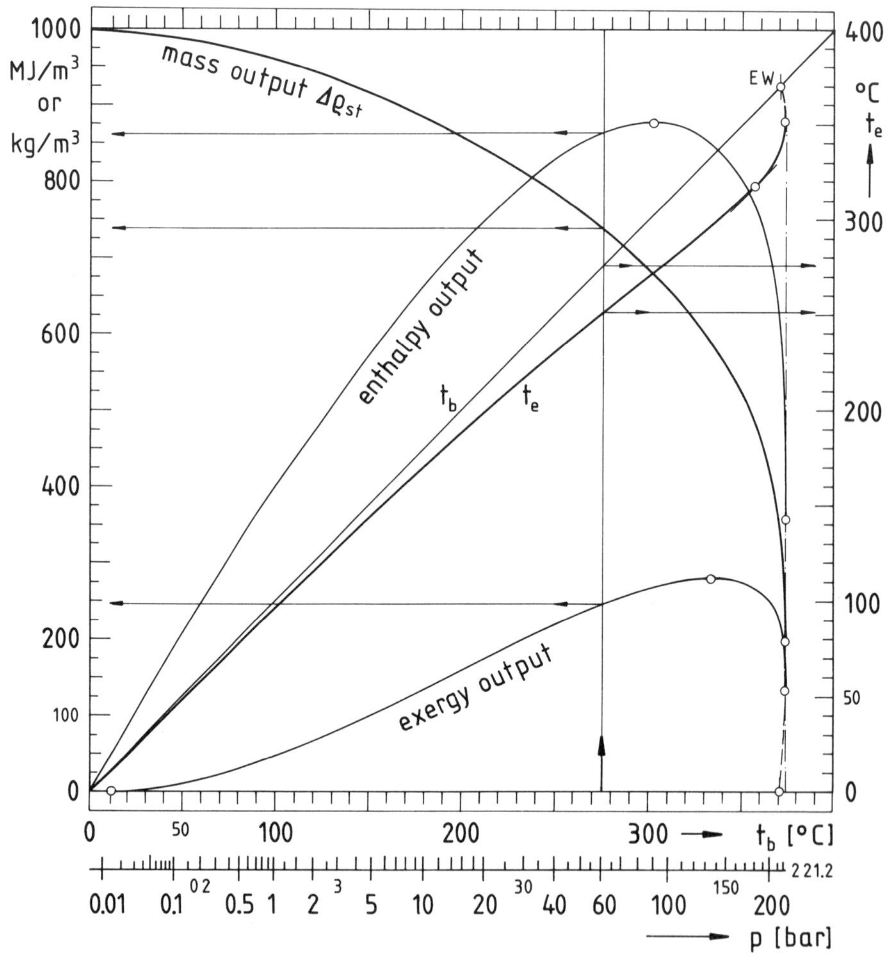

Fig. 2.6. Mass, enthalpy and exergy output of an expansion-type accumulator, between fully charged ($y = 0$) and fully discharged ($y = 1.0$) condition
t_b temperature at the begin of discharge, t_e temperature at the end of hot water discharge
Example:
Begin of discharge: $p_b = 60$ bar ($t_b = 275°C$), End of discharge: $t_e = 251°C$ ($p_e = 40.5$ bar)
Mass output = 738 kg/m³, Energy output = 862 MJ/m³, Exergy output = 245 MJ/m³

Fig. 2.6 shows the mass, enthalpy and exergy output, and the discharge pressure or temperature versus the charge pressure for a complete hot water discharge.

An interesting case–although one that is of little practical value–is the water discharge from pressurized steam storage vessels at high pressures, between the critical point "*CR*" (221.2 bar) and point "*EW*" (end of water discharge in Figs. 2.3, 2.4 and 2.5). In this region, more water is condensed than discharged; that is, the water level is rising during the first part of discharge. In "*EW*" there is equilibrium between the discharged and condensed water.

The location of point *EW* may be found from Eq. (2.6) by putting

$$h_{in} = 0, \qquad dQ = 0, \qquad h_{ex} = h', \qquad u_{st} = u''. \qquad (2.27–2.30)$$

From Eq. (2.7) it follows that

$$dm_{st}(h' - u'') = m_{st}\, du''. \tag{2.31}$$

Together with a volume balance

$$m_{st}v'' = V_{st} \tag{2.32}$$

or in differentiated form

$$dm_{st} = -\frac{V_{st}}{v''^2}\, dv'' \tag{2.33}$$

gives

$$dm_{st}v'' = -m_{st}\, dv''. \tag{2.34}$$

From a comparison with Eq. (2.31), we finally obtain the condition of state for *EW*

$$\frac{du''}{h' - u''} = -\frac{dv''}{v''}. \tag{2.35}$$

This condition is met at approximately 212 bar (370°C).

2.2.4. Indirect Sliding-pressure Storage

The storage medium is contained in a closed pressure vessel (indirect thermal storage, constant mass, and constant volume). Charging and discharging is achieved by transferring heat to and from a heat exchange surface–(6) and (9) of Fig. 2.2–heating and cooling the storage medium. If water and steam are in thermal equilibrium, then Eq. (2.20) may be employed, with $m_{st,b} = m_{st}$ = constant and $\Delta m_{ex} = 0$. The discharged heat ΔQ_{ex} thus becomes

$$\begin{aligned}\Delta Q_{ex} &= m_{st}(j_{st,b} - j_{st,e})\\ &\quad + V_{st}\left[\left(\frac{u'' - u'}{v'' - v'}\right)_b - \left(\frac{u'' - u'}{v'' - v'}\right)_e\right]\\ &\quad + \Delta Q.\end{aligned} \tag{2.36}$$

ΔQ is the net heat added (for instance, heat from the vessel wall minus heat losses). In practice, however, thermal equilibrium between steam and water volume will not prevail during a fast discharge operation. In this case, the steam cushion rather acts as a pressurizer and the system is similar to the constant-pressure system described below.

2.3. Storage of Sensible Heat in Pressurized (Subcooled) Liquids

Fig. 2.7 shows the storage in pressurized liquids. By maintaining a pressure above saturation pressure ("pressurization") or a temperature below saturation temperature ("subcooling"), boiling of the liquid storage medium is prevented. Normally, the system is kept at constant pressure. Pressurization may be carried out by the atmosphere (for hot water storage below 100°C or for high-boiling liquids), by pressurized inert gas, or by a two-phase pressurizer maintaining the system pressure in a steam cushion by evaporating some liquid or condensing some steam. The storage capacity is provided by changing only the sensible part of the enthalpy; this type of storage is therefore based on liquid phase.

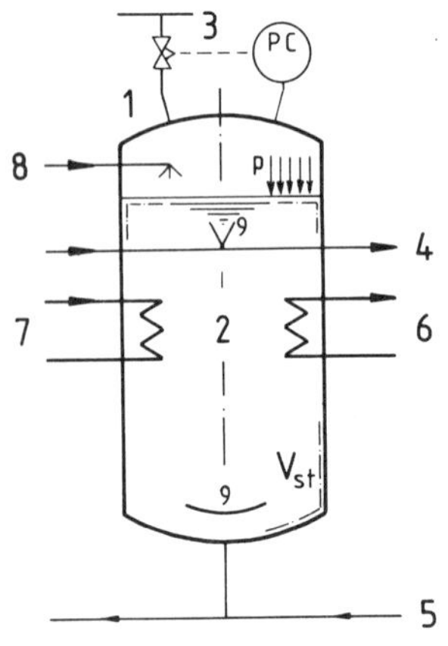

Fig. 2.7. Constant pressure accumulator
1 vessel, *2* liquid storage medium, *3* pressurizing system, *4* upper discharge/charge line, *5* lower discharge/charge line, *6* discharge heat exchanger surface, *7* charge heat exchanger surface, *8* upper charge line for (hot) water, *9* internals

Possible applications of this type of storage include processes with heat transmission by means of subcooled liquid or process cycles of which one step consists of heating up the working medium, such as feed water in steam power stations. So far, hot water storage in district heating systems, storage for feed water heating in steam power stations, and storage of hot oil or molten salts in solar power stations–where heat transfer medium and storage medium are identical–have been realized in practice.

2.3.1. Storage Capacity

For a liquid storage system under constant pressure, $dp = 0$. Eqs. (2.5) and (2.6) apply. With

$$V_{st} = m_{st}v \tag{2.37}$$

$$dV_{st} = dm_{st}v + m_{st}\,dv \tag{2.38}$$

the storage equation becomes

$$h_{in}\,dm_{in} + dQ - h_{ex}\,dm_{ex} = d(hm)_{st}. \tag{2.39}$$

Enthalpy input plus heat input minus enthalpy output yields enthalpy change of the storage medium. This is so regardless of whether the enthalpy change is caused by change of specific enthalpy h_{st}, or of storage mass m_{st}, or of both quantities. System borders and, therefore, the storage volume V_{st} are variable. Thus, volumetric work is performed during contraction (discharge) by either the ambient or an external vessel which is not part of the system considered. For the most part, discharge enthalpy h_{ex} remains constant, and a stepwise assessment is therefore not necessary.

If the storage mass is constant, then Eq. (2.39) further reduces to

$$dQ = m_{st}\,dh. \tag{2.40}$$

If only sensible heat is used, the enthalpy change may be written as

$$dh = c_p\,dT \tag{2.41}$$

$$dQ = m_{st}c_p\,dT = V_{st}\rho_{st}c_p\,dT \tag{2.42}$$

ρ_{st} density of storage medium.

The assessment of specific exergy storage capacity is based on the exergy content of heat which is a function of ambient temperature T_{am} [2.1]

$$\Delta E_{12} = \Delta Q_{12} - T_{am}\int_1^2 \frac{dh}{T}. \tag{2.43}$$

For the stored heat ΔQ_{12}, Eq. (2.40) applies, or with constant m_{st} and c_p and with the temperature swing $T_2 - T_1$

$$\Delta Q_{12} = Q_2 - Q_1 = m_{st}c_p(T_2 - T_1) \tag{2.40a}$$

$$\frac{\Delta E_{12}}{m_{st}c_pT_{am}} = [T_2/T_{am} - \ln T_2/T_{am}] - [T_1/T_{am} - \ln T_1/T_{am}]. \tag{2.44}$$

The dimensionless exergy storage function of sensible heat storage according to Eq. (2.44) is shown in Fig. 2.8 against the relative storage temperature based on ambient temperature. The dimensionless heat storage capacity

$$\frac{\Delta Q_{12}}{m_{st}c_pT_{am}} = T_2/T_{am} - T_1/T_{am} = (T_2 - T_1)/T_{am} \tag{2.45}$$

is represented in this diagram as a straight line under an angle of 45°.

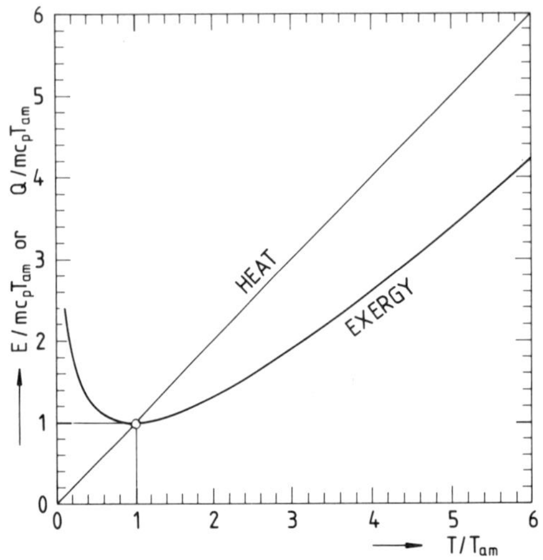

Fig. 2.8. Dimensionless exergy function and heat storage capacity versus dimensionless storage temperature of sensible heat storage

The diagram is used in a manner similar to the earlier discharge diagrams (see Figs. 2.3, 2.4 and 2.5). Energy content or exergy function are determined for the upper (T_2) and lower (T_1) condition. Heat output is found as difference ΔQ_{12} on the line "heat", and exergy output ΔE_{12} on the line "exergy". The exergy line starts at $T/T_{am} = 1$ horizontally; its slope increases with increasing T/T_{am} and approaches 1. It is easily shown that the slope is (in the range $T/T_{am} \geqslant 1$) equal to the Carnot factor

$$\left(\frac{dE}{m_{st} c_p T_{am}}\right) \Big/ dT \doteq 1 - T_{am}/T. \tag{2.46}$$

The shape of the exergy line in the region of $T/T_{am} < 1$ is remarkable. In going from 1 to 0, exergy increases and becomes infinite at 0. This means that exergy storage capacity at very low temperatures is large. In this region, discharge is linked to a temperature rise; the environment does not act as heat sink but rather as heat source, whereas the cold store acts as heat sink. It actually turns out that at $T_{st} \doteq 0.203\ T_{am}$ heat and exergy storage capacity are equal, and below that temperature exergy storage capacity is even larger than heat storage capacity.

The medium and high temperature range (but not the low temperature range) has been used for power storage thus far. One of the reasons for this is that the use of low temperatures requires thermodynamic storage (heat pump storage) with special working fluids and relatively costly machinery.

2.3.2. *Types*

The following types and flow sheets are possible (see Fig. 2.9):

(a) *Hot and Cold Vessel* (Fig. 2.9A)

Hot (charged) and cold (discharged) storage media are kept in separate vessels, each being designed for the total mass of storage medium. The cold vessel is to be designed for the lower temperature (and possibly pressure) and may be smaller because of the smaller specific volume (higher density). Transient thermal stresses and exergy losses by mixing are avoided.

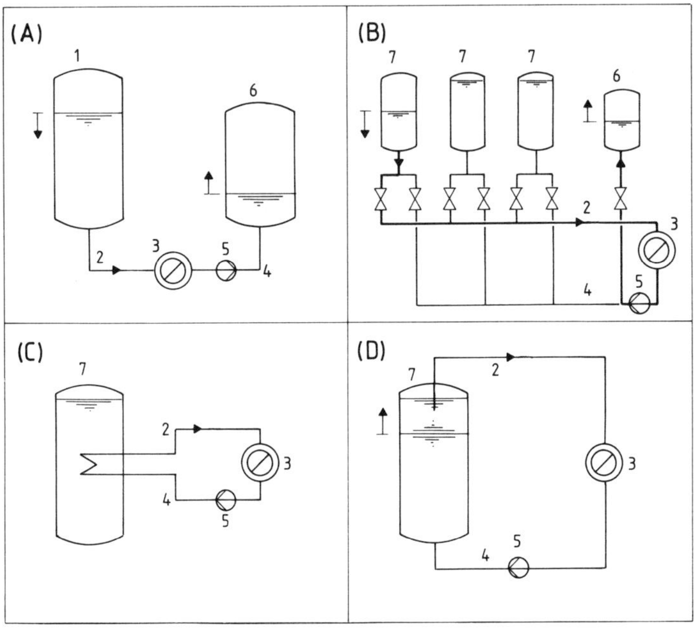

Fig. 2.9. Types and flow sheets of sensible heat storage in pressurized liquids
A hot and cold vessel, *B* farm concept, *C* sliding-temperature type, *D* displacement storage (thermocline)
1 hot vessel, *2* hot line, *3* consumer, *4* cold line, *5* pump for discharge, *6* cold vessel, *7* hot/cold vessel

(b) *Farm Concept* (Fig. 2.9B)

This type is similar to that in (a). There are n vessels. In the charged condition, the vessels No. 1 to No. $(n-1)$ are filled with hot liquid, while No. n is empty. During discharge, hot liquid is taken from No. $(n-1)$, and after cooling is stored in No. n, and so on until vessels Nos. 2 to n are filled with cold liquid and No. 1 is empty. There is less vessel volume required than with the hot and cold vessel concept in (a). But at least $n-1$ vessels are to be designed for the upper temperature (and pressure). Transient thermal stresses do occur but should present no major problem at low design pressures and wall thickness.

(c) *Sliding-temperature Type* (Fig. 2.9C)

The temperature of the liquid is uniformly increased during charging and decreased during discharging. There may be a heat exchanger in the vessel or the liquid may be pumped around.

(d) *Displacement Storage—"Thermocline"* (Fig. 2.9D)

Hot and cold liquid are in one vessel, preferably a vertical cylindrical one. The hot (less dense) liquid is in the upper part, the cold is in the lower part ("Thermocline"). Because of the heat conduction of water and the vessel wall and because of turbulences during charging and discharging, a temperature transmission zone (a mixing zone) will develop which leads to exergy losses but which also reduces transient thermal stresses in the vessel wall.

Turbulence is reduced and, in particular, proper distribution is enforced upon the incoming liquid by horizontal discs located near the entrance. Heat conduction in the water content may be reduced by floating thermally insulated discs or, more commonly, by slim vertical vessels. Nevertheless, this type is not well suited to low frequencies of charge and discharge.

2.3.3. Storage Media

There is a larger number of storage media suited for thermal storage of subcooled liquids than for liquids at saturation (as dealt with in Section 2.2). Data on some more important liquid storage media are presented in Table 2.1. Liquids with a boiling temperature (at atmospheric pressure) above 100°C are usually called "high boiling". The working temperature is, in any case, between melting temperature and boiling temperature. For liquids with high viscosity at low temperatures (such as high temperature oils), the minimum working temperature may be considerably above their melting temperature, and it may be given by the limits of pumping power.

Table 2.1 shows that volumetric heat capacity—the product of specific heat capacity c_p and density ρ—is high for water. However, water requires a pressure vessel at temperatures above 100°C. An interesting proposal [2.12] concerns the use of aqueous solutions of, for instance, sodium hydroxide (NaOH), permitting temperatures of 130°C, and possibly 150°C, without pressure vessel.

HITEC is the brand name of an eutectic mixture of molten salts with rather high volumetric heat capacity. As in the case of liquid metals, provisions have to be made to avoid freezing.

2.4. Sensible Heat Storage in Solids

The storage medium is a solid body that is heated up and cooled down without phase change. The sensible part of enthalpy determines its storage capacity. The equations of Section 2.3.1 apply here (with the specific heat capacity c instead of c_p).

Table 2.1. *Thermal Data of Liquid Storage Media*

Medium/Composition[1]	Temperature °C				Density		Specific Heat Capacity[2] c	Volumetric Heat Capacity[3] ρc		Vessel or Liner Material[4]
	Melting Temp. t_f	Lower Operation Temp. t_l	Upper Operation Temp. t_u	Boiling Temp. t_b	$\rho(t_l)$ kg/m³	$\rho(t_u)$	kJ/kg K	kJ/m³ K	kWh/m³ K	
(a) *Water* [2.8]										
H_2O (1 bar)	0	0	100	100	1000	958	4.19	4010	1.12	LA[a]
H_2O (10 bar)	0	0	180	180	1000	881	4.19	3690	1.03	LA[a]
H_2O (100 bar)	0	0	311	311	1005	688	4.29	2950	0.82	LA[a]
(b) *Mixtures* [2.10, 2.11]										
50 H_2O/50 Ethylenglykol	-36	0	100	–	1075	1015	3.48	3530	0.98	–
Ethylenglykol ($C_6H_6O_2$)	-13	0	100	197	1130	–	2.5	2700	0.75	–
(c) *Aqueous Solution*										
50 NaOH/50 H_2O	–	20[d]	140	140	1550	–	3.27	5060	1.4	–
(d) *Heat Transfer Oil*										
Petrochemical (average)	-20	–	–	300	900	750	2.4	1800	0.5	LA[b]
Synthetic (range)	-70/ -10	–	200/350	230/410	1000	750	2.1	1600	0.44	LA[b]
(e) *Molten Salt* [2.18]										
53 KNO_3/40 $NaNO_2$/7 $NaNO_3$	142	–	450	–	1970	1725	1.3	2300	0.64	LA[c]
(f) *Liquid Metals* [2.13]										
Na	98	–	–	880	925	750	1.26	945	0.26	AU
K	63	–	–	760	830	660	0.77	508	0.14	–

[1] Composition in percent by weight.
[2] At $t_m = (t_l + t_u)/2$.
[3] $\rho = \mathrm{Min}[\rho(t_l), (t_u)]$.
[4] LA Low alloy steel.
AU Austenitic steel.
[a] Degassed feedwater.
[b] For higher temperatures.
[c] Below 450°C.
[d] Limit of solubility.

2.4.1. Types

In principle, all of the types mentioned in Section 2.3.2 for liquid storage media are possible for solid storage media also. However, there is an additional distinction to be made for solid media: That is, whether the solid bodies themselves transfer the heat out of the containment vessel (pebble bed systems; see 2.3.2(a) and (b)) or whether these remain in the vessel and the heat is transferred by means of a fluid (liquid or gas; see 2.3.2(c) and (d)). In the latter case, the pressure of the heat transfer fluid will determine the design of the vessel.

If the storage medium is a solid body and is structured, then the "Thermocline" effect (see 2.3.2(d)) may be realized relatively easily. The vessel again gets a slim cylindrical shape with axial flow of the heat transfer medium (Fig. 2.10 A), short heat conduction paths, sufficient heat transfer area, and low thermal conductivity in the axial direction.

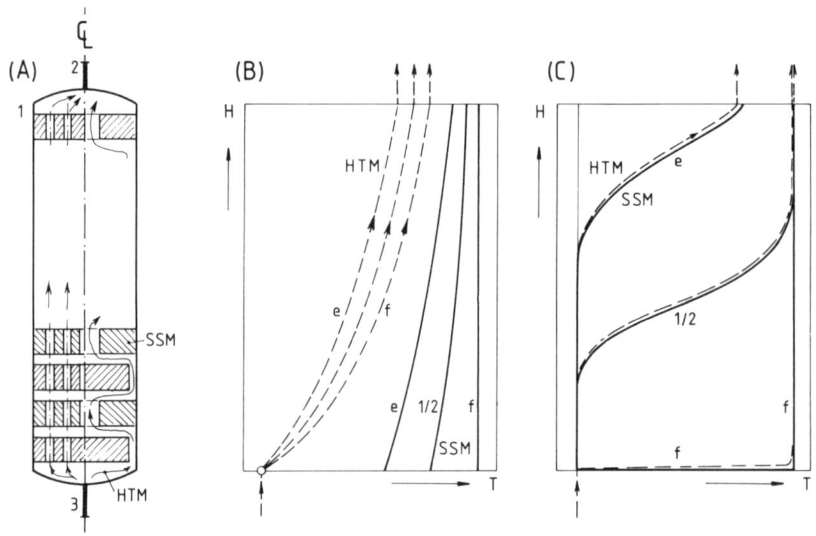

Fig. 2.10. Sensible heat accumulator with stationary solid storage medium
A arrangement of solid storage medium: *1* vessel, *2* upper discharge/charge line, *3* lower discharge/charge line, *SSM* stationary solid storage medium, *HTM* heat transfer medium; *B* temperature profile during discharge–poor heat transfer and high flow rate, *C* temperature profile during discharge–good heat transfer and low flow rate, *f* fully charged, 1/2 half charged, *e* empty

The temperature distribution during charging and discharging versus vessel height is shown in Fig. 2.10 B and C for two cases. Case B applies if the heat transfer area and the heat transfer coefficient are low relative to the flow rate (Cowper); case C applies if they are high (pebble bed). In both cases there will be axial heat transfer, thus causing a further flattening of the temperature lines.

In case C a temperature transition zone will develop and will move up and down during charging and discharging in much the same way as the mixing zone in the pure liquid displacement storage vessel. The outlet temperature will stay practically constant almost until the end of discharge. The heat capacity of the heat transfer fluid that is contained in the vessel is, of course, to be added to the heat capacity of the solid bodies. The fact that the heat transfer fluid always has to flow through the whole system and is subject to pressure losses along the whole path while the heat transfer takes place only in the rather narrow transition zone may be regarded as a disadvantage of this system.

2.4.2. Solid Storage Media

Some thermal properties of solid storage media are presented in Table 2.2. Volumetric heat capacities are generally high as are the possible temperature swings (which are hardly used to the full extent).

Table 2.2. *Thermal Data of Solid Storage Media*

Medium	Melting Temperature	Density	Specific Heat Capacity	Volumetric Heat Capacity		Thermal Conductivity
	t_f	ρ[1]	c	ρc		λ
	°C	kg/m^3	kJ/kgK	kJ/m^{3}K	kWh/m^{3}K	W/mK
(a) *Metals* [2.16]						
Steel (low alloy)	1510	7850	0.46	3615	1.00	50
Cast Iron	1150–1300	7200	0.54[2]	3920[2]	1.09	42[2]
Copper	1083	8960	0.39	3190	0.97	395
Aluminum	660	2700	0.92	2490	0.69	200
(b) *Non-Metals*						
Fireclay	–	2100–2600	1.0	2350	0.65	1.0–1.5
Al_2O_3 (90%)	1700[3]	3000	1.0	3000	0.83	2.5
MgO (90%)	1700[3]	3000	1.0	3000	0.83	4.5–6
Feolite	–	–	–	3700	1.03	5
Rock	–	1900–2600	0.8–0.9	1600–2300	0.44–0.64	1.5–5.0

[1] At room temperature.
[2] Mean value between 20 and 300°C (depends on temperature and composition), according to Siempelkamp.
[3] Refractoriness under load.

Cast iron shows the highest heat capacity of the metals. Of the non-metals, fireclay has a fair heat capacity but a poor thermal conductivity. High alumina (Al_2O_3) and high magnesia (MgO) have high heat capacities. However, magnesia is better with respect to heat conduction and cost and is therefore considered to be a good thermal storage medium. Feolite, a sintered iron oxide in brick form [2.14, 2.15], is another good solid storage medium.

When comparing the heat capacity of solid with that of liquid storage media, one must account for the void of the solid system (gap volume; the volume of channels for the heat transfer medium) on the one hand, and, on the other hand, for the heat capacity of the heat transfer fluid contained in the vessel. In other words: The effective heat capacity is the weighted mean of the heat capacities of solid body and heat transfer medium. Number, shape and size of the channels may be determined by the nature of the solid (for example, in the case of cast iron, narrower and more tortuous channels will be possible than with refractory material). In the case of a rock bed, this void volume is to be taken into account by using the bulk density instead of the apparent density of the stone. Randomly packed beds with frequent temperature cycling may be subject to settling and radial expansion, possibly leading to liner or vessel failures.

2.5. Latent Heat Storage

The storage capacity of latent heat is based not on temperature change but on phase change of the storage medium at constant temperature.

The phase change from the liquid to the gaseous phase involves the largest latent heat. However, the volumetric storage capacity of the vapor phase is rather low. Therefore, this type of latent heat storage has not been used or suggested. For the most part, latent heat storage is understood to mean the storage of melting heat which involves only small volume changes. Sometimes, the solid-liquid phase change is combined with a solid-solid phase change at a temperature somewhat below the melting point.

It is often suggested that use be made of the sensible heat of the liquid and/or solid phase in addition to the latent heat. This will indeed increase the heat capacity but it will also remove the advantage of heat supply at constant temperature.

Contrary to all of the previously discussed storage systems which are more or less state of the art, latent heat storage systems are still in the development stage. The incentives are: high heat capacity, constant temperature and low pressures. The problems are: heat transfer between storage medium and working medium, cost of heat exchangers, suitable temperature levels and controlled atmosphere.

2.5.1. Storage Capacity

Latent heat storage is a constant pressure, constant mass system. Therefore, Eq. (2.40) applies. The enthalpy increase of a solid/liquid system from $T_1 < T_s$ to $T_2 > T_f$ is

$$\Delta h = c_{p,s}(T_f - T_1) + \Delta h_f + c_{p,l}(T_2 - T_f) \tag{2.47}$$

with

- $c_{p,s}$ specific heat capacity, solid phase (p = constant)
- Δh_f enthalpy of phase change (melting at T_f = constant)
- $c_{p,l}$ specific heat capacity, liquid phase (p = constant).

The first term of Eq. (2.47) is the sensible part of the solid phase, while the second is the latent heat, and the third is the sensible part of the liquid phase.

Exergetic storage capacity is determined in analogy to Section 2.3.1

$$\frac{\Delta E}{m_{st}} = c_{p,s}(T_f - T_1 - T_{am} \ln T_f/T_1) + \Delta h_f(1 - T_{am}/T_f)$$
$$+ c_{p,l}(T_2 - T_f - T_{am} \ln T_2/T_1). \tag{2.48}$$

Knowledge of Δh_f is important. For pure substances, there are approximations [2.17] as follows:

– for organic substances:

$$\Delta h_f \doteq 0.7\, T_f \tag{2.49}$$

$$\rho\, \Delta h_f \doteq \rho\, 0.7\, T_f \tag{2.50}$$

where Δh_f is in kJ/kg, T_s in K, ρ in kg/m^3 and $\rho\, \Delta h_f$ in kJ/m^3;

– for anorganic substances:

$$\Delta h_f \doteq (24/M) T_f \tag{2.51}$$

$$\rho\, \Delta h_f \doteq (24\rho/M) T_f \tag{2.52}$$

where M is the molecular mass in kmol and ρ is in kg/m^3. For most substances, ρ/M is of the order of 40.

For instance, water (ice) has a melting storage capacity of about 335 kJ/kg ((24/18)273 = 364 kJ/kg according to the approximation). This is as much as a sensible hot water system with a temperature swing of about 335/4.19 = 80 K.

At higher temperatures, capacities are higher. By choosing binary mixtures (in particular eutectics) or mixtures with more than two components, the melting temperature is reduced, without appreciable reduction of the phase-change enthalpy.

2.5.2. Storage Media

In chosing a latent heat storage medium, the following properties are important:

(1) low cost;
(2) high phase-change enthalpy; high density;
(3) suitable melting temperature;
(4) high thermal conductivity in the solid and liquid phase (heat exchanger!);
(5) high heat capacity in the solid and liquid phase (if sensible heat is also used);
(6) no tendency to segregation, thermal stability;
(7) no subcooling during freezing, no supersaturation during melting;
(8) low thermal expansion; low volume change during melting;
(9) either no or low corrosivity (this is to allow use to be made of inexpensive vessel material);
(10) low risk (no poisonous vapors, no dangerous reactions with working medium or heat transfer medium).

Table 2.3. *Thermal Data for Latent Heat Storage*

Medium/Composition[1]	Melting Temperature	Density[2]		Heat of Fusion	Volumetric Heat Capacity[3]		Specific Heat Capacity		Thermal Conductivity	Vessel or Liner Material[4]	Safety Hazard[5]	Ref.
	t_f	ρ_s	ρ_l	Δh_f	$\rho\ \Delta h_f$		c_s	c_l	λ_s			
	°C	kg/m^3		kJ/kg	MJ/m^3	kWh/m^3	kJ/kgK		W/m K			
(a) *Pure Substances*												
H_2O	0	917	1000	335	307	85	2.1	4.2	2.2	P	0	
$LiNO_3$	252	2310	1776	530	941	261	2.025	2.04	1.35	AU	–	
NaOH	318	2130	1780	160	285	78	2.01	2.09	0.92	LA[b]	2[c]	2.18
LiOH	471	1425	1385	1080	1500	416	3.3	3.9	1.3	LA[b]	2	2.18
Al	660	2560	2370	400	948	263	0.92	–	200	–	–	2.21
Li_2CO_3	726	2114	1810	607	1100	305	–	–	1.45	AU	–	2.19
$Na_2B_4O_7$	740	2300	2630	530	1220	339	1.75	1.77	–	–	–	
LiF	848	–	–	1050	1800	500	–	–	–	AU[d]	–	2.20
(b) *Binary Mixtures*												
$48NaCl/52MgCl_2$	450	2225[a]	1610	431	694	193	0.92	1.0	–	AU	2	2.18
$33NaCl/67CaCl_2$	500	2160[a]	1900	282	536	149	0.84	1.0	–	AU	1	2.18
$35Li_2CO_3/65K_2CO_3$	505	2265[a]	1960	345	676	188	1.34	1.76		AU	2	2.18
88Al/12Si	579	2553	2445	515	1259	350	1.49	1.27	high	–	–	2.21
$67LiF/33MgF_2$	741	–	–	900	2000	556	–	–	–	AU	–	2.20
$67NaF/33MgF_2$	832	2690	2190	618	1353	376	1.42	1.38	4–12	AU	3[d]	2.18
(c) *Ternary Mixtures*												
$24.5NaCl/20.5KCl/55MgCl_2$	385–393	–	1800	410	738	205	–	–	1.0	–	–	2.23
$32Li_2CO_3/35K_2CO_3/33Na_2CO_3$	397	2300	2140	277	593	165	1.68	1.63	–	AU	2	2.18

1 Composition in percent by weight.
2 Near melting temperature.
3 $\rho = \mathrm{Min}[\rho_s, \rho_l]$.
4 LA Low alloy steel, AU Austenitic steel, P Plastic.
5 Toxic hazard rating code: 0 = none, 1 = slight, 2 = moderate, 3 = high.

a At room temperature.
b If suitable corrosion inhibitors are used.
c Reacts with H_2O.
d When strongly heated, toxis fumes are emitted.

(a) *Pure Substances*

Part (a) of Table 2.3 shows properties of pure substances such as t_f, Δh_f, $\rho\ \Delta h_f$ (where ρ refers to the lower of either the solid or the liquid state), thermal conductivity, etc. From the specific heat capacities in the liquid and solid phase, the suitability of the medium for sensible heat storage may be judged. The thermal conductivity is a measure for the heat transfer, particularly during discharge. Density change and corrosivity determine design and material of the storage vessel. Comments on safety aspects are also given.

Of the substances included–water/ice, salts and metals–lithium fluoride (LiF) shows the highest heat of melting; it also is the most expensive of the substances. Lithium hydroxide (LiOH) is almost as good in terms of latent heat, its melting temperature being suitable to many processes. However, it is also expensive. All of the salts show low thermal conductivities and large volume changes during melting. The metals show high thermal conductivities, and some of them show high energy densities.

(b) *Binary Systems*

There are several incentives for binary systems in latent heat storage:

- Melting points may be in more suitable ranges.
- High energy densities may be achieved even at low melting temperature.
- Expensive substances with good storage properties may be combined with inexpensive ones, the heat capacity remaining almost the same.

A condition for the use of binary systems is that it melts and freezes like a homogeneous pure substance. This will occur for two special compositions, the *eutectic* and the *dystectic* composition.

(1) The *eutectic composition* is represented as the lowest point in the melting diagram. If the mixture is slightly off the eutectic composition, then a strong increase of the melting point normally takes place.

(2) The *dystectic composition* is a mixture with a certain stoichiometric composition that creates a mixing phase (a chemical compound). It therefore behaves almost like a pure substance.

In the melting diagram, such a mixture is represented by an intermediate maximum. Fig. 2.11 shows such a dystectic point at the 1:1 molar composition of the Li_2CO_3-K_2CO_3 system (that is 35% Li_2CO_3/65% K_2CO_3 by weight). This compound has an eutectic with K_2CO_3 (28/72 by weight) and another one with $LiCO_3$ (47/53 by weight) [2.18].

If the mixture is neither eutectic nor dystectic or if the dystectic is metastable, the melting process is said to be incongruent, that is, the solid phase decays into a liquid and another solid phase. The different compositions and, therefore, the different densities of liquid and solid appearing and disappearing phases may lead to segregation. A typical example for this highly unwanted effect is Glauber's salt ($Na_2SO_4.10H_2O$) which has been proposed for storing low temperature solar heat for space heating purposes on account of its low cost, its melting temperature of 32°C, and its high heat of melting [2.22].

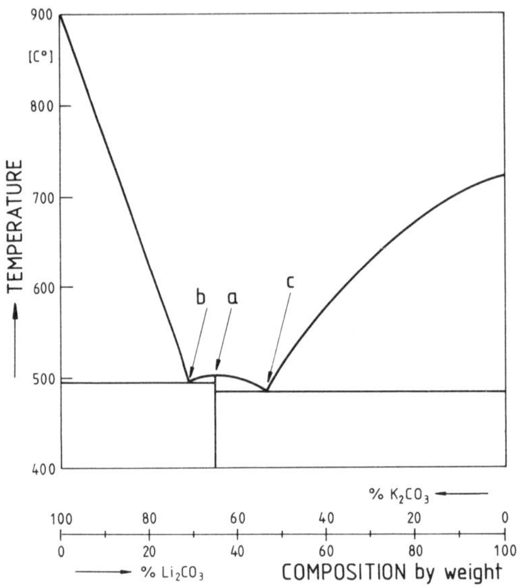

Fig. 2.11. Phase diagram of the Li_2CO_3-K_2CO_3 system [2.18]
a dystectic point, *b, c* eutectic points

Properties of some eutectic and dystectic mixtures are given in part (b) of Table 2.3. The chlorides are inexpensive but show a high volume change, while the carbonate mixtures show a low volume change but are more expensive.

(c) *Ternary Mixtures*

Eutectic mixtures may also consist of three or more components. In part (c) of Table 2.3, a chloride and a carbonate mixture are listed which show thermal properties that are as good as the binary mixtures but possess a lower melting point and lower cost.

2.5.3. Charge and Discharge Equipment

As was mentioned before, with latent heat storage it is not as easy to transfer the heat to the storage medium during charging and from the storage medium during discharging as it is with sensible heat storage systems. The reasons for this are that convection is prevented in the solid state, thermal conductivity is generally low, and the volume is changing. The following suggestions for overcoming this problem have been made:

(1) Encapsulation of the storage medium (for example, in tubes that are in immediate contact with the working medium steam/water). Fig. 2.12 shows this concept, the pressure vessel being a prestressed cast iron vessel [2.24].

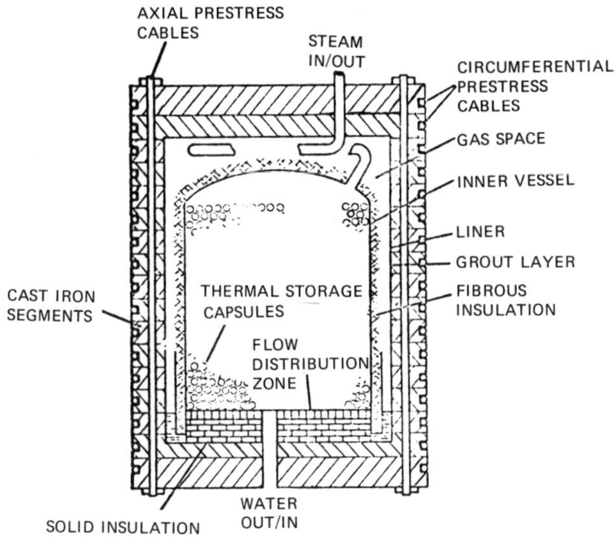

Fig. 2.12a. Prestressed cast iron vessel for latent heat storage [2.24]

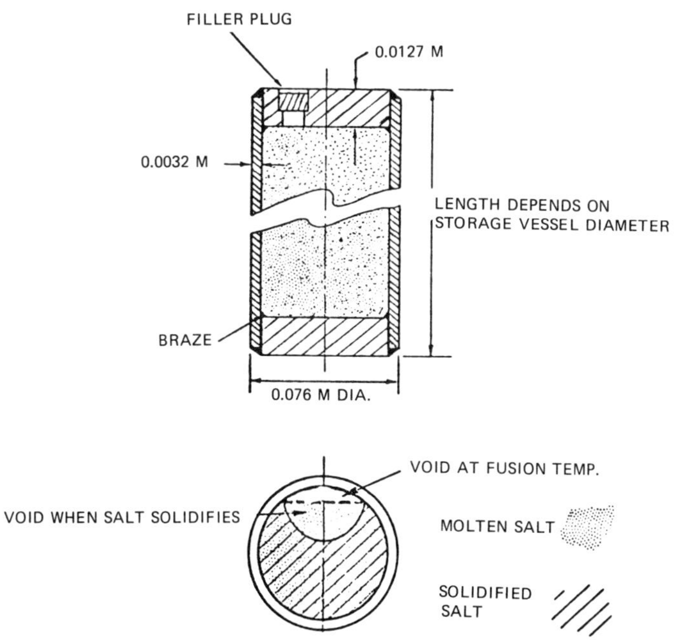

Fig. 2.12b. Detail to Fig. 2.12a: Cylindrical capsule

(2) Encapsulation of the storage medium in small individual containers which are placed in a containment vessel at almost atmospheric pressure, a separate heat transfer medium being employed for the transfer of heat to and from the charging/discharging heat exchanger by evaporation/condensation similar to a heat pipe. This "energy storage/boiler tank concept" is shown in Fig. 2.13. The proposed storage medium is an eutectic salt mixture; the containment vessel is filled with *m*-terphenyl (boiling point 365°C) and storage elements; the liquid heat transfer medium and a charging heat exchanger is in its lower part. Spray nozzles are located in the upper parts; and the discharge heat exchanger [2.23] is placed above these.

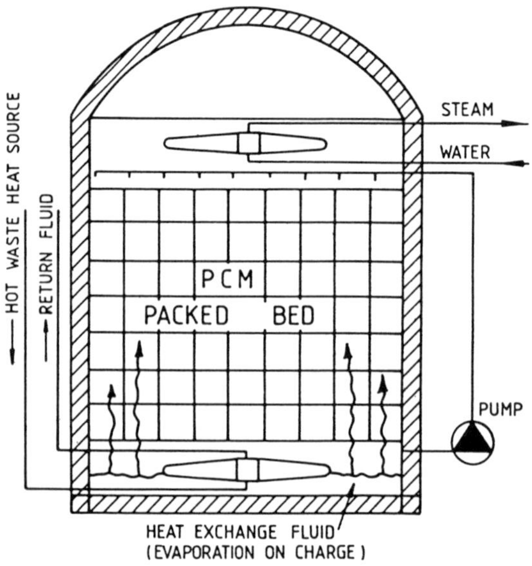

Fig. 2.13. Energy storage boiler tank [2.15, 2.23]

(3) Storage medium on the shell side of the discharge heat exchanger, multiple rotating or translatory scrappers or large rotating tubes with fixed scrappers.
(4) Storage medium on the shell side of densely packed finned [2.25] or unfinned tubes.
(5) Mechanically generated vibration (for example, ultrasonic) of the storage medium which is not completely frozen.
(6) Flexible tubes of the discharge heat exchanger that are subject to pressure pulses.
(7) Heat transfer via liquid metal which is in direct contact with the storage medium.

All of these proposed solutions of the heat transfer problem of latent heat storage are presently still in the engineering or the testing stage.

2.6. Pressurized Gas Storage

Contrary to the steam cushion storage system, the pressurized gas storage system uses (single-phase) gas as storage medium. There is no phase separation; the discharged gas has the same state as the remaining stored gas. The storage system consists of a pressure vessel with or (what is more common) without thermal insulation. If wet air or gas is used, a water drain may be required.

Industrial applications range from steel pressure vessels of up to several cubic meters behind air compressor plants for equalizing supply pressure or for enabling on-off operation of the compressor to underground pressurized air storage systems with volumes of more than 100 000 cubic meters for peak load coverage by means of a gas turbine.

The assessment begins with Eq. (2.4). Because $dm_{in} = 0$ and $V_{st} = 0$, Eq. (2.4) reduces to

$$dQ - (u + pv)_{ex}\, dm_{ex} = du\, m_{st} + u\, dm_{st}. \tag{2.53}$$

But since

$$u_{ex} = u_{st} = u \tag{2.54}$$

and

$$\rho_{st} V_{st} = m_{st} \tag{2.55}$$

it further reduces to

$$dQ/V_{st} + (pv)\, d\rho_{st} = \rho_{st}\, du. \tag{2.56}$$

If the storage medium is assumed to be an ideal gas, then we have

$$pv = p/\rho = (c_p - c_v)T \tag{2.57}$$

$$du = c_v\, dT \tag{2.58}$$

$$\kappa = c_p/c_v \tag{2.59}$$

$$dQ/V_{st} + (c_p - c_v)T\, d\rho_{st} = c_v \rho_{st}\, dT. \tag{2.60}$$

For the *adiabatic* case ($dQ = 0$), Eq. (2.60) may be written as

$$(\rho/\rho_0)_{st}^{\kappa - 1} = T/T_0 \tag{2.61}$$

where the index "0" refers to the initial conditions. Because of Eq. (2.57), T may be eliminated and the *discharge equation* will then read

$$(\rho/\rho_0) = (p/p_0)^{1/\kappa}. \tag{2.62}$$

Because of Eq. (2.61), the discharge temperature becomes

$$T/T_0 = (p/p_0)^{(\kappa - 1)/\kappa}. \tag{2.63}$$

Fig. 2.14 shows the specific storage capacity of a pressurized air storage system (air was considered to be an ideal two-atomic gas, with $\kappa = 1.4$). The diagram was made dimensionless by relating all values to the initial state. Because of Eq. (2.57) which may be written as

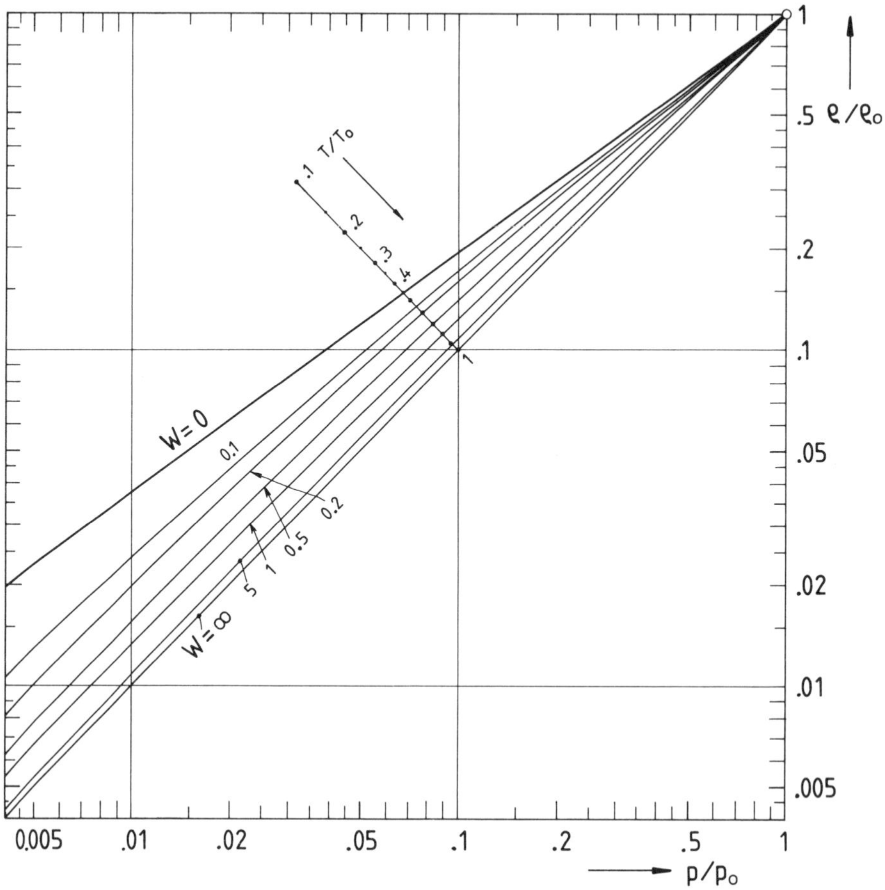

Fig. 2.14. Storage capacity of compressed air storage
Example for $W = 5$:
Begin of discharge: $p_b = 100$ bar, $t_b = 25°C$, $T_b = 273 + 25 = 298$ K, $\rho_b = 118$ kg/m^3
End of discharge: $p_e = 15$ bar, $p/p_0 = 15/100 = 0.15$, $\rho/\rho_0 = 0.16$, $T/T_0 = 0.95$
$\rho_e = 0.16(118) = 18.9$ kg/m^3
Mass output $= 118 - 18.9 = 99.1$ kg/m^3, $T_e = 0.95(298) = 283$ K, $t_e = 283 - 273 = 10°C$

$$\log \rho = \log p - \log T \tag{2.64}$$

the temperature may also be taken from the double logarithmic presentation of ρ and p. The temperatures appear as oblique straight lines having a 45° slope. The use of the diagram starts by choosing the pressure ratio (final to initial pressure)

$$\epsilon = p/p_0 < 1. \tag{2.65}$$

After taking ρ/ρ_0 from the diagram, the discharged amount of gas (air) per cubic meter storage volume is calculated from the known initial density ρ_0

$$\rho_{ex} = \rho_0 - \rho = \rho_0(1 - \rho/\rho_0). \tag{2.66}$$

An approximately *isothermal* discharge may be achieved either by a very slow discharge rate and heat exchange with ambient or by a separate latent storage system. The heat to be stored follows from Eq. (2.60) for $dT = 0$

$$dQ_{T=\text{const}}/V_{st} = -(c_p - c_v)T\,d\rho_{st} = -(\kappa - 1)c_v T\,d\rho_{st}. \tag{2.67}$$

The discharge equation follows from Eq. (2.56) and T = constant

$$\rho/\rho_0 = p/p_0. \tag{2.68}$$

A comparison to Eq. (2.62) shows the increase of storage capacity by isothermal versus adiabatic discharge.

In most practical cases, the discharge is neither adiabatic nor isothermal. A certain amount of heat will be added—for example, by the heat capacity of the pressure vessel—but it will not be enough to avoid the temperature drop completely. In such cases, storage capacity has to be assessed—similar to the procedure with sliding-pressure storage—by solving the storage equation (2.53) stepwise. If no time lag of vessel wall temperature is assumed, the heat added becomes

$$\Delta Q = m_{fe} c_{fe} (T_b - T_e) \tag{2.69}$$

m_{fe} mass of vessel wall
c_{fe} specific heat capacity of vessel wall
T_b initial temperature
T_e final temperature.

Using Eqs. (2.8) and (2.10) and applying the equation for an ideal gas, we see that the storage equation becomes

$$\Delta\rho c_v[\kappa T_m - T_e] = \rho_b c_v (T_b - T_e) - \Delta Q/V. \tag{2.70}$$

With $T_m = (T_b + T_e)/2$, and on basis of Eqs. (2.69) and (2.70), the storage equation finally becomes

$$\Delta\rho/\rho_b = -\left(\frac{1 - T_e/T_b}{(\kappa/2) - (1 - \kappa/2)T_e/T_b}\right)\left(1 + \frac{m_{fe}c_{fe}}{V\rho_b c_v}\right). \tag{2.71a}$$

In the special case of air ($\kappa = 1.4$) we obtain

$$\Delta\rho/\rho_b = -\left(\frac{1 - T_e/T_b}{0.7 - 0.3T_e/T_b}\right)\left(1 + \frac{m_{fe}c_{fe}}{V\rho_b c_v}\right). \tag{2.71b}$$

The expression $W = (m_{fe}c_{fe}/V\rho_b c_v)$ is the heat capacity ratio of vessel wall to initial gas content at constant volume. For a small vessel and air at high pressures, it may be around $W = 5$. Fig. 2.14 shows lines for various values of W; the higher W happens to be, the closer the isothermal case is approached.

If additional heat storage capacity is provided (for example, solid storage or latent storage), W may be increased further. In this way, not only is the storage capacity increased but also the temperature drop during discharge is minimized, which may be of advantage for the process and will avoid undue thermal stresses.

Fig. 2.15 shows the exergy density of pressurized air storage, for the case in which the initial storage temperature and the discharge temperature are equal to the ambient temperature and the expansion within the storage vessel as well as in the engine is isothermal. It should be noted that exergy density is independent of ambient temperature: A high ambient temperature will lower the density in the store but will increase the enthalpy of the stored air. These two effects thus cancel each other out. Exergy density increases more than proportionally to storage pressure. High pressures (above 50 bar) have therefore been proposed for pressurized air storage.

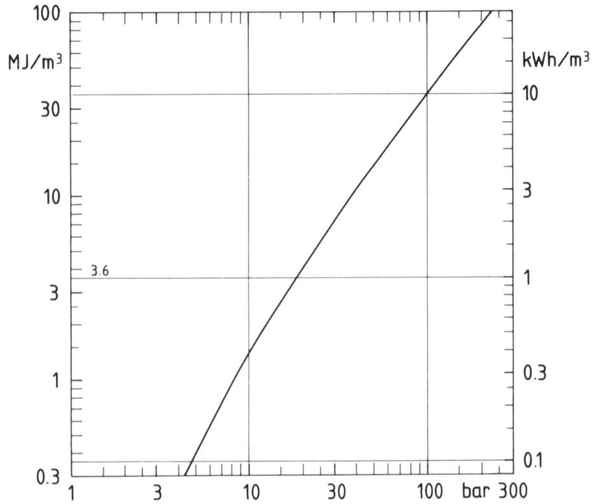

Fig. 2.15. Exergy density of compressed air storage

Its heat storage capacity being negligible, pressurized air storage is not applied to heat storage but rather to the power storage of the gas turbine power plant. There, a furnace chamber or a heat exchanger is situated between store and gas turbine. For capacity reasons, a low storage temperature is mostly intended.

2.7. Other Thermal Storage Systems

Apart from temperature change and/or phase change, the change of binding energy between two or more storage media may also be used for thermal energy storage, provided that the physical or chemical reaction is a reversible one.

High energy densities are possible if the reaction partners can be stored in the liquid state. In such systems, discharge by parasitic heat losses because of incomplete thermal insulation such as in sensible or latent heat storage may be completely avoided by separating and separately storing the two or more media that are produced during the charging. Therefore, these storage systems are well suited for long-time storage.

According to the kind of binding, there are two types to be distinguished: sorption heat storage and thermochemical heat storage.

2.7.1. Sorption Heat Storage

Some systems are based on the following reaction pattern:

$$AB \text{ (solid or liquid)} + \Delta Q_V \rightleftarrows A \text{ (solid or liquid)} + B \text{ (gaseous)}.$$

The substance A may be an element, a component or a mixture [2.26]. The heat of reaction (at atmospheric pressure) is about

$$\Delta h_s = (125 \text{ to } 150) T_s / M_B \tag{2.72}$$

Δh_s sorption enthalpy (kJ/kg)
T_s sorption temperature (°K)
M_B molecular mass of B (kmol/kg).

Therefore it is larger than the heat of melting or the heat of evaporation. However, the mass of A is also to be accounted for.

Fig. 2.16 shows four basic flow sheets of sorption heat storage systems.

According to Fig. 2.16 (*A*), heat is added to the sorbent material (for instance, $CaCl_2$), the desorbed substance (NH_3) is led to a gas tank working at constant or sliding pressure. Cooling of the gas will increase its storage capacity. Energy density is low.

If the gas may be condensed at or near ambient temperature (and at proper pressure), a flow sheet according to Fig. 2.16 B is possible, and little storage volume is required. Heat of condensation will be transferred to ambient (for example, by means of a cooling tower). At the discharge stage, the liquid is again vaporized by ambient heat. Some exergy loss is unavoidable. This storage arrangement is identical to an intermittently working absorption refrigerator in which the absorber medium (2) is alternatively exposed to heating and cooling. In the heating mode, heat is disposed at the condenser (6); in the cooling mode, the system pressure falls, and the condensate starts to boil in the evaporator (7), where cold is made available. It should be noted that the continuously working absorption refrigerators, chillers, or heat pumps do not possess these inherent storage features. If thermal storage is required, storage vessels for both the rich and poor solution must be added.

In the flow sheet corresponding to Fig. 2.16 C, liquid storage is replaced by sorption storage containing another absorber medium which is capable of absorbing and desorbing gas at ambient temperature (the resorption principle).

Fig. 2.16 D shows an interesting combination of sorption storage and heat engine. When the discharge valve is opened, steam flows from the boiler via a steam engine to the sorption store, heat is released and added to the boiler, thus sustaining the generation of steam. Charging may be carried out by a steam compressor, according to the principle of the compression heat pump. Or it may be

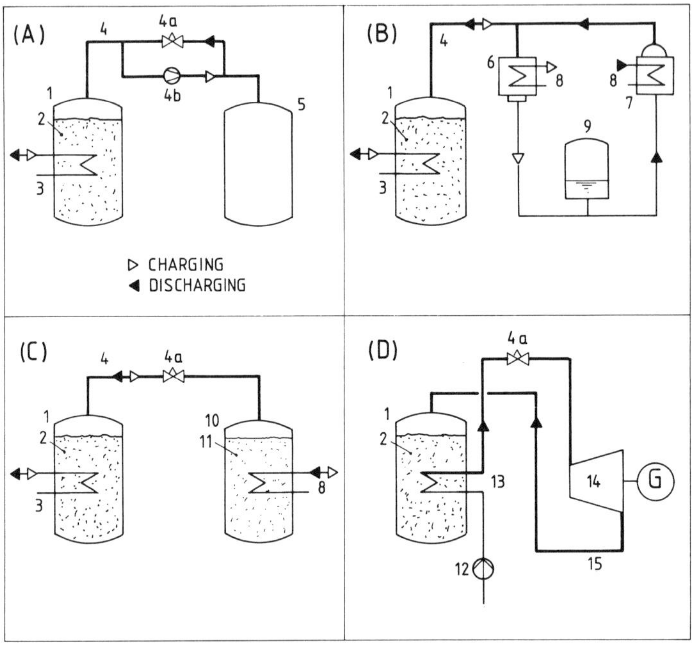

Fig. 2.16. Types and flow sheets of sorption heat storage
A with gas storage, *B* with condensate storage, *C* resorption cycle, *D* with boiler and steam turbine
1 absorber vessel, *2* absorber medium, *3* discharge/charge heat exchanger surface, *4* vapor line, *4a* throttling valve, *4b* compressor, *5* gas storage, *6* condenser, *7* evaporator, *8* low exergy heat source, *9* condensate storage, *10* resorber vessel, *11* resorption medium, *12* feed pump, *13* boiler, *14* steam turbine, *15* exhaust steam line

carried out by the principle of the absorption heat pump by heating the sorption store with a high temperature heat source and piping the desorbed and condensed gas to the boiler. A similar system, with NaOH as absorber and H_2O as working medium, was used in the locomotive of Honigmann (see Sections 1.4 and 8.3.2).

Table 2.4 shows the thermal properties and energy densities of some substances suitable for sorption heat storage. Although the heat of sorption is favourable for the pair $CaCl_2.8NH_3/4NH_3$, energy density is low, because of the low bulk density of the absorbing salt. Energy densities are listed for

(a) storage of the desorbed substance in the gaseous state according to Fig. 2.16 A,
(b) storage of the desorbed substance in the liquid state according to Fig. 2.16 B.

The energy densities are rather low for (a), even at higher pressures. For (b), the energy densities are higher but charging and discharging losses are also higher.

The sorbent material has to be in loose bulk form in order to enable storage and a sufficiently quick reaction. Heat conduction of the bulk material is even

Table 2.4. *Thermal Data of Thermochemical Storage Media*

Reaction		Storage									Volumetric Storage Density		Ref.
Kind[1]	Equation[2]	Condition of Reaction		Component	Condition of Phase[2]	Pressure	Temperature	Density	Weight Unloaded/Loaded	Required Volume			
		Pressure bar	Temperature °C			bar	°C	kg/m^3	kg/kg	m^3/kg	MJ/m^3	kWh/m^3	
S	$CaCl_2 . 8NH_3(s)$ + 745 kJ/kg	1	47–32	$CaCl_2 . 8NH_3$	s	1	20	550[3]	1–0.725	0.0018			2.26
	= $CaCl_2 . 4NH_3(s) + 4NH_3(g)$	10	90	NH_3	l[4]	10	20	600	0–0.275	0.0005[4]			
		30	100	NH_3	g[5]	10	90	5.7	0–0.275	0.0482[5]			
										0.0023[4]	327	91[4]	
										0.0501[5]	15	4[5]	
S	$MgCO_3(s)$ + 1200 kJ/kg	1	427–327	$MgCO_3$	s	1	20	1500[6]	1–0.48	0.0007			2.26
D	= $MgO(s) + CO_2(g)$			CO_2	l	74	31	465	0–0.52	0.0011			2.27
										0.0018	672	187	
S	$Ca(OH)_2(s)$ + 1415 kJ/kg	1	572–402	$Ca(OH)_2$	s	1	20	1115[6]	1–0.757	0.0009			2.28
D	= $CaO(s) + H_2O(g)$			H_2O	l	1	20	1000	0–243	0.0002			2.29
										0.0011	1241	345	
C	$SO_3(g)$ + 1235 kJ/kg	1	520–960	SO_3	l	1	45	1900	1–0	0.0005			2.30
	= $SO_2(g) + 1/2\ O_2(g)$			SO_2	l	6.3	40	1320	0–0.8	0.0006			2.31
				O_2	g	100	20	130	0–0.2	0.0015			
										0.0027	462	128	
C	$CH_4(g) + H_2O(g)$ +	1	480–1195	CH_4	g	100	20	67	0.47–0	0.0070			2.32
	+ 6053 kJ/kg			H_2O	l	1	20	1000	0.53–0	0.0005			
	= $CO(g) + 3H_2(g)$			CO	g	100	20	117	0–0.82	0.0070			
				H_2	g	100	20	8	0–0.18	0.0225			
										0.0371	163	45	
D	$NH_4HSO_4(l)$ + 2011 kJ/kg = $NH_3(l) + H_2O(l) + SO_3(l)$	–	927–427	NH_4HSO_4	l	–	–	–	1–0	–	3071	858	2.29

1 S = sorption, C = catalytic, D = decomposition.
2 s = solid, l = liquid, g = gaseous.
3 Bulk density.
4 Storage of NH_3 in liquid condition.
5 Storage of NH_3 in gaseous condition.
6 Void volume 50%.

lower than that of salts. Therefore, the problems of heat transfer during charging and discharging will be even more difficult to solve.

Another possibility is the use of carbonates such as $MgCO_3$, releasing CO_2 when heated. Use of the metal hydrides presents still another possibility. Here too, heat of sorption is high but energy density is low if the hydrogen is stored in the gaseous state. And liquifaction is difficult. However, with a flow sheet according to Fig. 2.16 C, another metal hydride (for instance, a Fe-Ti system) may, at ambient temperature, take up the hydrogen released during charging. Metal hydrides have also been proposed for storing hydrogen as a fuel for combustion engines and fuel cells, in particular for cars [2.33]. Zeolites with steam as desorbed substance have also been proposed [2.26].

Prototypes were built (locomotive of Honigmann, vehicle of Isshiki, see Chapter 8). Further development is still at the stage of selecting working pairs and studying the reaction kinetics.

2.7.2. Thermochemical Storage

Thermochemical storage consists of storing thermal energy as binding energy of reversible chemical reactions (sometimes thermochemical storage is defined to include sorption heat storage also). The reaction may take place with or without a catalyst. After the reaction, the reaction partners are to be separated and stored separately.

Table 2.4 also shows some reactions that have been proposed for thermochemical storage. The charging reactions take place from the left-hand side to the right-hand side. The heat of reaction refers to 1 kg total mass. The reaction temperature is the so-called "reversing temperature" where the reaction coefficient is 1. Reaction products are gaseous. If at least one of the reaction products is stored in the liquid state, then the energy densities will be favourable. However, during the condensation (charging), heat of evaporation (which normally is not used and reduces storage efficiency) is set free.

The dissociation of SO_3 has been proposed for future solar power stations with gas turbines [2.30]. Energy density is favourable in spite of the high pressure storage of oxygen. This is particularly so if the SO_2 that is produced during the charging operation is fed to the emptied SO_3 containment vessels (similar to Fig. 2.9 B). However, problems concerned with the lifetime of the catalyst and with the materials are expected.

The so-called Eva-Adam reaction (methane and steam) was proposed for heat transport and storage of future helium-cooled High Temperature Reactors [2.32]. For storage only, the system is not very attractive because reaction products are gaseous and energy density is therefore low.

The dissociation of NH_4HSO_4 has a very high energy density, since all products may be stored in the liquid state. The salt has a low melting point (144°C), it is relatively inexpensive, and the reaction products NH_3, SO_3 and H_2O are easily separated, condensed, stored and evaporated. Although this kind of thermochemical storage may have some future, its development is not yet completed.

2.8. Comparison of Storage Capacity

2.8.1. Energy Density

Fig. 2.17 shows the energy densities of thermal storage systems. The following void volumes have been used:

12.5% for liquid storage media;
25% for solid storage with metal matrix;
37.5% for solid storage with non-metallic matrix and for phase-change storage media.

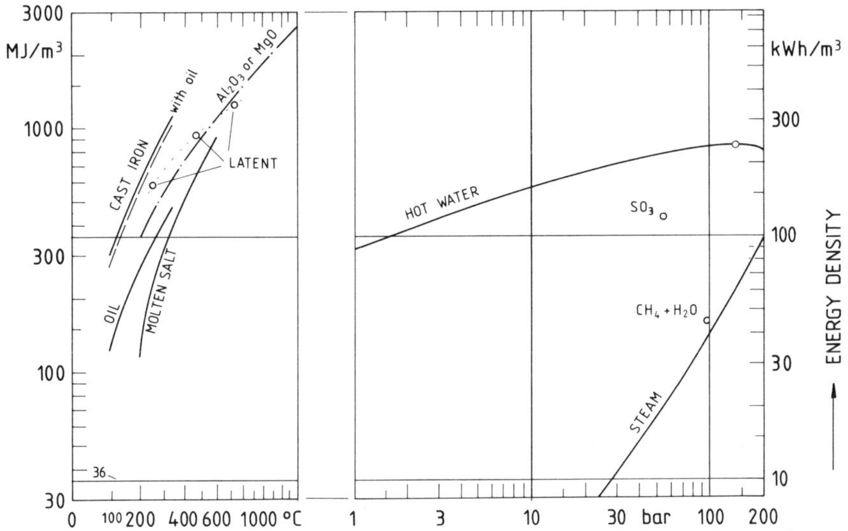

Fig. 2.17. Energy densities of storage media

(a) *Unpressurized Systems*

Storage systems with low (below atmospheric) saturation pressures are shown in the left-hand diagram of Fig. 2.17. The containment may, therefore, be unpressurized. Energy density for a discharge from upper temperature to ambient temperature is plotted against the upper temperature. Thermo-oils may be used up to 350°C and molten salts above that temperature.

Latent heat storage shows a higher energy density than sensible heat storage (although only the heat of melting and no sensible part is taken into account). An envelope of favourable media was drawn. However, some of the media show even higher energy densities (for example, lithium salts and the eutectic fluoride mixture).

Of the solid storage media, cast iron shows the highest energy density (again based on enthalpy difference from upper to ambient temperature).

(b) *Pressurized Systems*

The right-hand diagram of Fig. 2.17 shows the energy density of media with saturation pressures above atmospheric (necessitating a pressure vessel) versus storage pressure. There is little difference between hot water storage and sliding-pressure steam storage. The slopes of the lines are strongly degressive. For heat storage as such there is little incentive for high pressures.

Sorption heat storage requires at least one pressure vessel and thermochemical heat storage several pressure vessels at different pressures if the reaction partners are stored in the liquid state. The diagram is based on the average pressure $p = (\Sigma\, pV)/\Sigma\, V$.

2.8.2. *Exergy Density*

The exergy density of the same storage system as in Fig. 2.17 is again plotted in Fig. 2.18 for the unpressurized systems versus temperature and for the pressurized systems versus pressure. An ambient temperature of 10°C (283°K) was assumed, this temperature corresponding to the average air temperature in Central Europe.

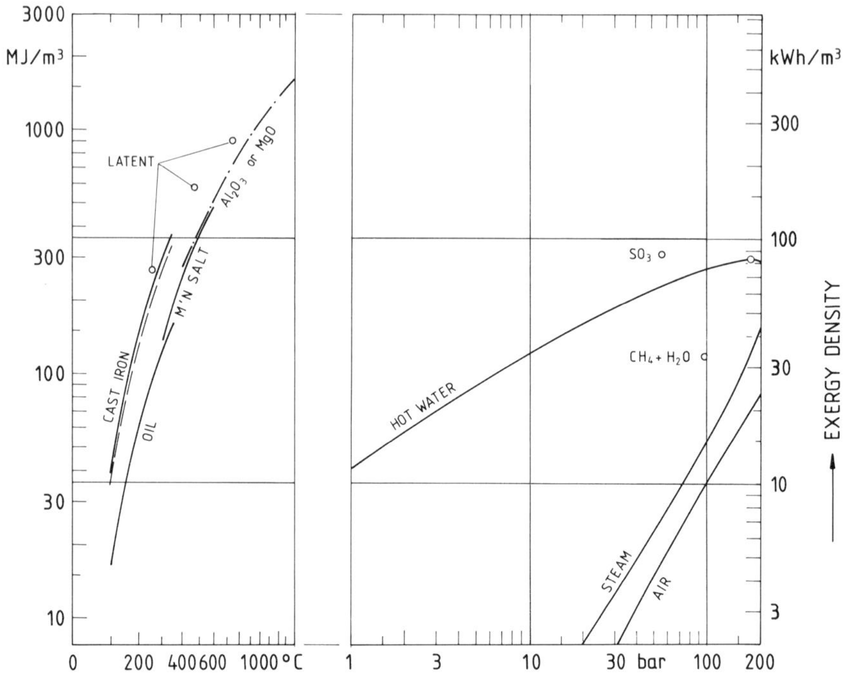

Fig. 2.18. Exergy densities of storage media

The latent heat storage systems show a relative gain since their constant discharge temperature means a higher efficiency than the decreasing temperature of sensible heat storage systems. Nevertheless, except for the lithium salts mentioned above, cast iron storage shows the highest exergy density above about 350°C. Molten salts also improve relative to the enthalpy density because they can only be cooled down to their freezing point, with correspondingly higher exergetic efficiency.

The line of hot water is still degressive, but it is steeper than in Fig. 2.17, so that there is an incentive to use high pressures (for example, up to 50 or 70 bar). Pressurized air is much below hot water.

2.9. Efficiency Factors of Heat Storage

The energy and exergy densities of the previous sections are theoretical upper limits. In practice, there will be stagnation losses (by heat dissipation and mixing) as well as losses during charging and discharging (by heat transfer, throttling and mixing).

These losses constitute the efficiency factor of the storage vessel ζ_{ac}, this factor being defined as the ratio of the discharged exergy to the charged exergy that is required to reconstitute the initial charged condition. In special cases (for example, if heat is stored without reference to power generation) the efficiency factor may be defined as enthalpy (energy) ratio. It should then be named calorific efficiency factor ν_{ac}.

2.9.1. Stagnation Losses

Stagnation losses occur during the time period between the charge and discharge processes as self-discharge. If they occur in the plant (for example, in pipes), they are normally counted as starting losses of the plant. Stagnation losses are different for sensible, latent and thermochemical storage.

(a) *Stagnation Losses of Sensible Heat Storage*

The storage enthalpy at temperature T based on ambient temperature T_{am}, with storage mass m_{st} and specific heat capacity (at constant pressure) c_p, is

$$\Delta Q = m_{st} c_p (T - T_{am}). \tag{2.73}$$

Because of incomplete thermal insulation, there is a heat flow to ambient

$$\dot{Q} = dQ/dz = - kA(T - T_{am}) \tag{2.74}$$

where

- k overall heat transmission coefficient
- A surface area
- z time.

From Eqs. (2.73) and (2.74), we have

$$T(z) - T_{am} = [T(z = 0) - T_{am}] \exp(- zkA/m_{st} c_p). \tag{2.75}$$

Storage temperature decreases, finally approaching ambient temperature. The calorific stagnation efficiency factor

$$\nu_{acst}(z) = \Delta Q(z)/\Delta Q(z=0) \tag{2.76}$$

therefore becomes

$$\nu_{acst}(z) = \exp(-zkA/m_{st}c_p) = \exp(-Z^*). \tag{2.77}$$

This is plotted in Fig. 2.19 against "dimensionless time"

$$Z^* = zkA/m_{st}c_p. \tag{2.78}$$

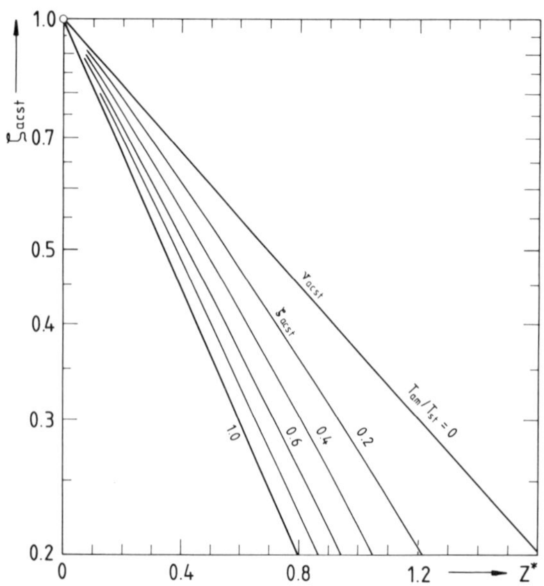

Fig. 2.19. Exergetic and calorific efficiency versus dimensionless standstill time

For the (exergetic) stagnation efficiency factor the Carnot factor is accounted for

$$\zeta_{acst}(z) = \frac{\Delta E(z)}{\Delta E(z=0)} = \left(\Delta Q\,\frac{T-T_{am}}{T}\right) \Big/ \left(\Delta Q\,\frac{T-T_{am}}{T}\right)_{z=0}. \tag{2.79}$$

Using Eqs. (2.75) and (2.77) we obtain

$$\zeta_{acst} = \frac{1}{1-(T_{am}/T)(1-\exp(Z^*))} \cdot \exp(-Z^*). \tag{2.80}$$

This is a function of the temperature ratio T/T_{am}. From Fig. 2.19 and Eq. (2.80) the following special cases are apparent:

– High temperature storage ($T_{am}/T_{st} \doteq 0$): For a very high upper temperature

T_u (uppermost line), the exergetic efficiency factor is equal to the calorific efficiency factor

$$\zeta_{acst} = \exp(-Z^*) = \nu_{acst}. \tag{2.81}$$

– Low temperature storage ($T_{am}/T_{st} \doteq 1$): For an upper temperature approaching T_{am} (lowest line), the exergetic efficiency factor is equal to the square of the calorific efficiency factor

$$\zeta_{acst} = \exp(-2Z^*) = \nu_{acst}^2. \tag{2.82}$$

Exergy losses by mixing of the hot and cold portion of the storage medium in displacement storage vessels do occur independently of external heat losses, and they are to be treated separately (see Chapter 4).

(b) *Stagnation Losses of Latent Heat Storage*

There will also be heat losses to ambient. However, the storage temperature remains constant by definition. Thus, the losses are independent of time and Eq. (2.74) applies. The Carnot factor is to be added to obtain the exergy loss; it too is time-independent. The exergetic and the calorific efficiency factors are equal.

(c) *Stagnation Losses of Thermochemical Storage*

The reverse reaction may be completely avoided by separate storage of the components. However, there may be heat losses of the components, constituting a start-up loss. Such a loss, however, will generally be of a negligible proportion.

2.9.2. Charge and Discharge Losses

Apart from some heat losses of the pipes, the charge and discharge losses are exergy losses only (without enthalpy losses). Some examples of this are: exergy losses in heat exchangers, by throttling as well as by mixing.

(a) *Exergy Losses in Heat Exchangers*

Exergy losses in heat exchangers are caused by finite temperature differ-

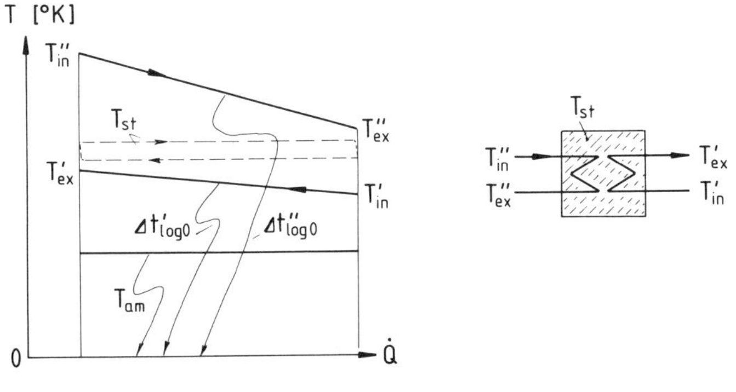

Fig. 2.20. T/Q-plot for an arrangement with one heat exchanger

ences and (usually much less important) by friction and by mixing of the flow paths of different temperatures.

Fig. 2.20 shows a temperature-enthalpy diagram with the temperatures of the primary (heat supply) and the secondary (heat demand) sides. Assuming that there are no enthalpy losses and that there is constant heat capacity, the two temperatures are represented by straight lines. With (′) referring to primary and (″) to secondary side

$$\dot{Q} = \dot{m}''c''(T''_{in} - T''_{ex}) = \dot{m}'c'(T'_{ex} - T'_{in}) \tag{2.83}$$

$$dQ = \dot{m}''c''\,dT'' = \dot{m}'c'\,dT'. \tag{2.84}$$

Using Eq. (2.43) for the exergy E of heat, the exergy loss is $E'' - E'$, and the exergetic efficiency of the heat exchangers becomes

$$\zeta = \dot{E}'/\dot{E}'' \tag{2.85}$$

with

$$\dot{E}'' = \dot{Q}\left[1 - T_{am}\frac{\ln T''_{in}/T''_{ex}}{T''_{in} - T''_{ex}}\right] \tag{2.86}$$

$$\dot{E}' = \dot{Q}\left[1 - T_{am}\frac{\ln T'_{ex}/T'_{in}}{T'_{ex} - T'_{in}}\right]. \tag{2.87}$$

Using the concept of "logarithmic temperature distance" (from absolute zero)

$$\Delta t_{\log 0}(T_1, T_2) = \frac{T_1 - T_2}{\ln T_1/T_2} \tag{2.88}$$

the exergetic efficiency finally becomes

$$\zeta_{HX} = \left(1 - \frac{T_{am}}{\Delta t_{\log 0}(T'_{ex}, T'_{in})}\right) \Bigg/ \left(1 - \frac{T_{am}}{\Delta t_{\log 0}(T''_{in}, T''_{ex})}\right). \tag{2.89}$$

ζ_{HX} approaches 1 only if $T'_{ex} = T''_{in}$ and $T'_{in} = T''_{ex}$ which is only possible for counterflow of two media with equal heat capacity ($c'\dot{m}' = c''\dot{m}''$) and infinite heat transfer area.

If one medium (say, the primary one) changes its phase ($T'_{in} = T'_{ex}$) and if $A = \infty$ and therefore $T'_{ex} = T''_{ex}$, then

$$\Delta t_{\log 0}(T_1, T_2 = T_1) = T_1 \tag{2.90}$$

and

$$\zeta_{HX} = \frac{1 - T_{am}/T''_{ex}}{1 - T_{am}/\Delta t_{\log 0}(T''_{in}, T''_{ex})}. \tag{2.91}$$

Sometimes stepwise heat exchange is employed to reduce these losses (see Fig. 2.21). Optimization yields the geometric mean as the intermediate temperature

$$T_Z = \sqrt{T''_{in}T''_{ex}}. \tag{2.92}$$

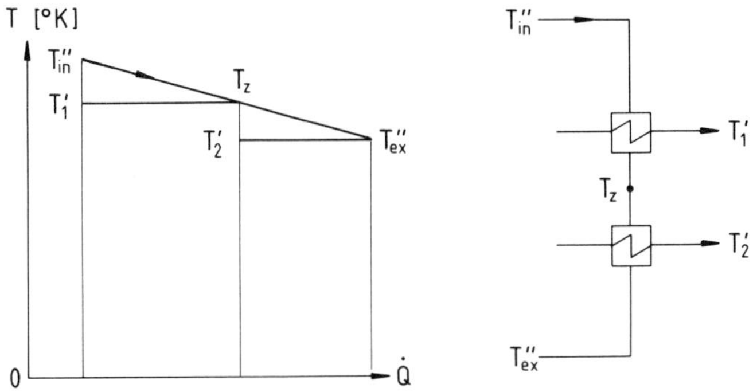

Fig. 2.21. *T/Q*-plot for an arrangement with two heat exchangers (phase change on the primary sides and infinite temperature approach in each heat exchanger is anticipated)

(b) *Exergy Losses of Throttling*

Throttling occurs as friction pressure drop in any flow through pipes and valves, etc. It is also used for load control and in flash evaporation (see Chapter 3).

The specific exergy of a flowing medium is [2.1, 2.2]

$$e = h - h_{am} - T_{am}(s - s_{am}) + c^2/2 + gz. \tag{2.93}$$

Neglecting the kinetic and the geodetic term and remembering that $h =$ constant during throttling, we have

$$e_2 - e_1 = - T_{am}(s_2 - s_1). \tag{2.94}$$

For ideal gases

$$s_2 - s_1 = c_p \ln(T_2/T_1) - (c_p - c_v)\ln(p_2/p_1) \tag{2.95}$$

and, furthermore, $T_2 = T_1$. Eqs. (2.94) and (2.95) therefore reduce to

$$s_2 - s_1 = -(c_p - c_v)\ln(p_2/p_1) \tag{2.96}$$

$$e_2 - e_1 = - T_{am}(c_p - c_v)\ln(p_1/p_2). \tag{2.97}$$

(c) *Exergy Losses Because of Mixing*

The exergy loss of adiabatic, isobaric mixing of mass flows of equal composition can be treated as a heat exchange with infinite temperature approach (see Fig. 2.22). If T_1 stands for the higher inlet temperature, T_2 for the lower inlet temperature and T_{mix} for the outlet temperature (which can easily be calculated from a heat balance), the exergetic efficiency becomes

$$\zeta = \frac{1 - T_{am}/\Delta t_{\log 0}(T_{\text{mix}}, T_2)}{1 - T_{am}/\Delta t_{\log 0}(T_1, T_2)}. \tag{2.98}$$

(d) *Charge and Discharge Losses of Thermochemical Storage*

Because of the finite reaction rate of the components, the charging temperature has to be higher than the discharging temperature. The heat of reaction is

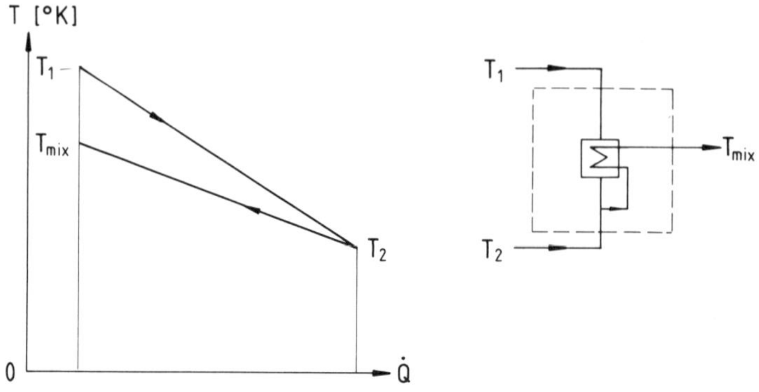

Fig. 2.22. Exergy loss due to mixing

practically the same in both directions. Therefore, the loss is a pure exergy loss and may be treated as a heat exchanger loss.

However, thermochemical storage permits an interesting way to keep the charge and discharge temperature at the same level (for example, by producing the required shift of equilibrium by a pressure change). Instead of a temperature loss there will be a loss caused by the necessary (parasitic) pumping power. This may, nevertheless, greatly alleviate the integration of the storage plant into the overall flow sheet.

2.9.3. Overall Storage Losses

The overall efficiency factor ζ_{ac} of the storage vessel, including its charge and discharge, consists of three terms: charging efficiency ζ_c, stagnation efficiency factor ζ_{acst} according to Section 2.9.1 and discharge efficiency ζ_d

$$\zeta_{ac} = \zeta_c \zeta_{acst} \zeta_d . \tag{2.99}$$

Charge and discharge losses are discussed in Section 2.9.2, they may each consist of an energy loss because of heat exchanger, throttling or mixing.

If the whole storage system is considered, further losses will occur, leading to the so-called turnaround efficiency of the storage system which is discussed in the following chapter.

3. Storage Systems

Apart from the storage vessel and its internals (see Chapter 4), a storage plant needs further external equipment to fulfil its function.

For heat storage (such as in process heat and district heating applications) pumps, heat exchangers, evaporators, valves and piping may be required for charging and discharging.

Even more extensive equipment may be necessary for power storage, and what might be called a "storage system" thus emerges. The store may be integrated into a power plant and charged from this power plant; or there will be a charging system with a separate energy converter such as a steam generator or a compressor. For discharge, there will be a power engine plus a generator and a heat sink. There will also be control and monitoring equipment.

3.1. Power Plant Storage Systems

3.1.1. Basic Principles

The basic principles of power plant storage systems are shown in Fig. 3.1. By analogy to hydraulic storage systems, the main classification is into flow storage and pumped storage systems.

The *flow storage system* gets its charge energy from the heat cycle of the power plant. In Fig. 3.1 A, the following flow storage systems are shown:

- compression energy storage (a), particularly for gas turbine cycles;
- feed water storage (b) for the steam process with regenerative feed water heating;
- primary cycle heat storage (c) for thermal power plants with separate heat transfer and working cycles such as is present in solar thermal power plants;
- working cycle heat storage (d) and (e) such as steam storage, superheat storage and gas turbine high temperature storage.

There are criteria for further classification of flow storage systems:

- integrated storage systems (with a base load plant capable of covering base load without the use of the storage systems) either with main turbine for base load and a separate peaking turbine, or

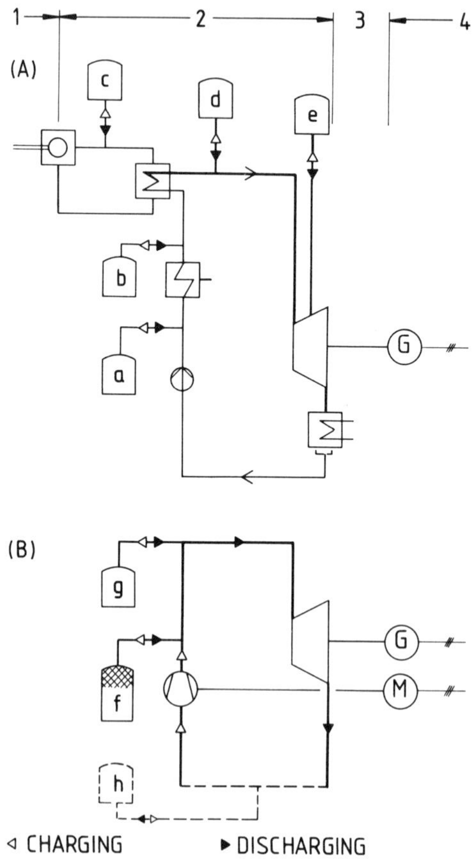

Fig. 3.1. Basic principles of power plant storage systems
A flow storage system, *B* pumped storage system, *1* primary energy, *2* secondary energy, *3* mechanical energy, *4* electrical energy

- with an overloadable main turbine that also covers peak load; and
- flow storage systems with separate energy converter (steam generator) and separate engine (peak load turbine).

A further distinction can be made between constant parameter storage (in which the storage medium is at the same place extracted from the process and reintroduced into the process from the store, "reversible storage") and variable parameter storage in which reintroduction occurs at a different place (downstream) or with considerably lower parameters.

Pumped storage systems are charged by electrical or mechanical energy. In Fig. 3.1 B the following pumped storage systems are shown:

- pneumatic storage (f) with and without separate storage of the heat of compression;
- heat pump storage (for example, with steam compressor), with upper store

(g) and lower store (h); one of them–for instance, the lower one–may be replaced by the ambient.

Pumped storage systems are usually charged from the electric grid and may therefore be placed anywhere in the grid. This means that they need not be integrated into the power plant. However, integrated pumped storage systems, charged by mechanical energy from a gas turbine or steam turbine, have also been proposed.

Normally, the upper store has parameters (pressure, temperature) above ambient, while the lower store has parameters near ambient or the environment itself may act as lower store (heat source storage system). Pumped storage systems may also be designed with the environment acting as upper store, while the lower store will have parameters much lower than ambient (heat sink storage system). In this case, heat for the process is taken from the environment whereas the lower store takes up the waste heat and becomes heated up [3.1]. As was pointed out in Section 2.3.1, the exergy density of such "negative exergy" systems may be rather high–theoretically infinite at absolute zero.

The classifications given in the above discussion are of practical significance for the assessment of storage efficiency and of charge and discharge power and speed, as well as for the selection of mode of operation and of performance control.

3.1.2. Flow Storage Systems in Steam Power Plants

On basis of the classifications in Fig. 3.1, we see that in steam plants direct storage by means of feed water (b) or by means of steam or hot water (d, e) is suitable. Fig. 3.2 shows the possibilities of combinations of type of storage and type of discharge. Charge equipment is not shown.

There are three types of storage [3.2]:

(A) displacement storage at constant pressure, with a transition zone between the hot and the cold part;
(B) expansion storage with almost constant pressure and hot water discharge from the storage vessel; and
(C) sliding-pressure storage (Ruths type), with internal steam generation and internal natural recirculation.

As regards the type of discharge, there are also three cases to be considered:

(1) steam generation by (external or internal) flash evaporation;
(2) cascading storage, a combination of (1) and (3);
(3) substitution of hot feed water (feed water storage).

Cases (C2) and (C3) do not exist. (B) and (C) may be combined; that is to say, steam may, in addition to hot water, be taken from an expansion storage vessel as is shown by the dotted lines in Fig. 3.2.

High storage efficiency may be obtained (at least, theoretically) by constant parameter storage systems. Feed water storage (A3) may be regarded as a constant parameter system if the feed water temperature is constant at the loads considered and the heat losses as well as the mixing losses in the storage vessel

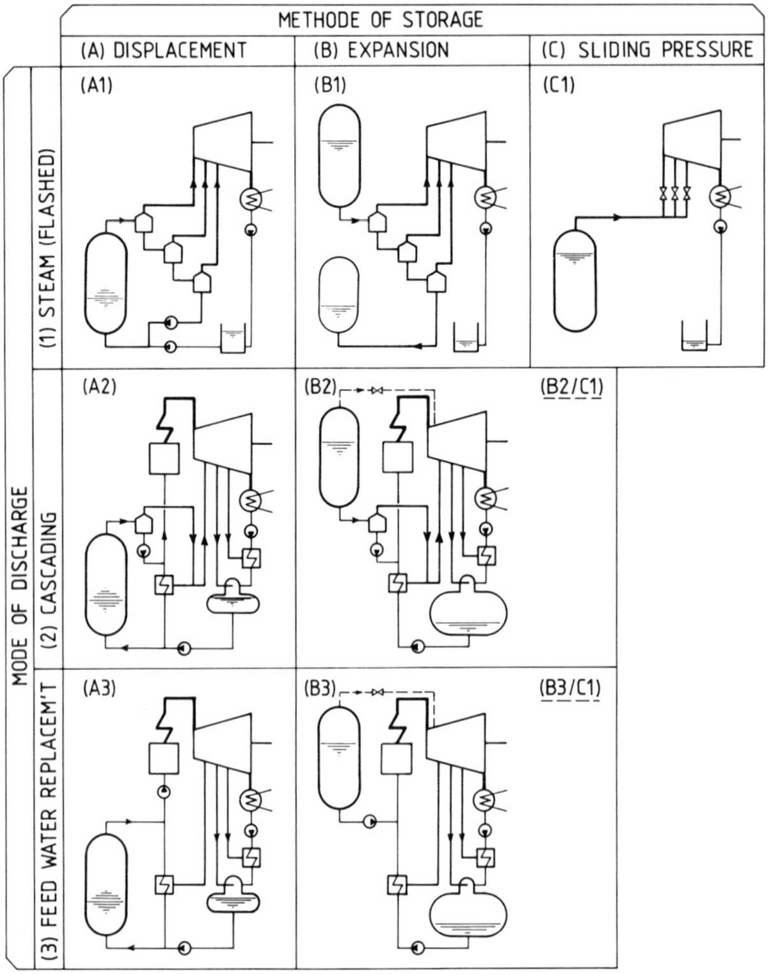

Fig. 3.2. Classification of pressurized hot water storage [3.2]

are low. On the other hand, cascading storage (2) adapts easily to changes of feed water temperature with changing load, since flashing takes place exactly down to the prevailing feed water temperature. Sliding-pressure storage (C1) becomes an almost constant parameter system if the pressure change is very low. However, this entails low specific steam output (and, therefore, high storage vessel volume, cost and heat loss).

3.1.3. Flow Storage Systems in Gas Turbine Power Plants

Although in steam power plants the working medium may be used as storage medium, this is not suitable in gas turbine power plants because of the low volumetric heat capacity of hot gas (even if pressurized). Solid storage has therefore been proposed for flow storage in gas turbine power plants, thus substituting the furnace chamber either partially or totally (Fig. 3.3).

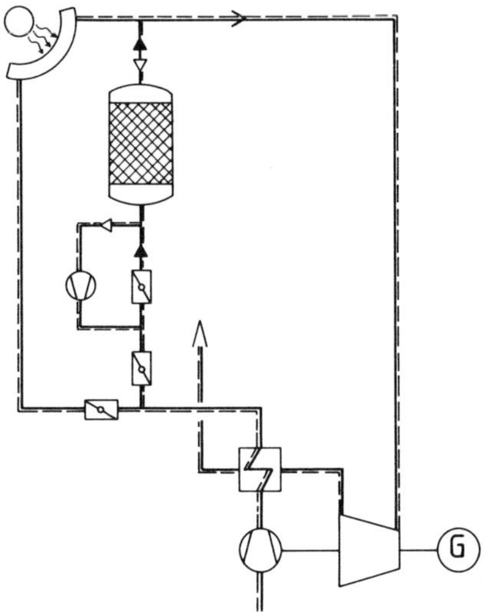

Fig. 3.3. Solar gas turbine power plant with solid heat storage

The storage vessel is to be designed for the gas pressure after the compressor, the pressurized hot gas acting as a heat transfer medium to the solid body (magnesia bricks or cast iron blocks arranged in a checkerwork arrangement). Such a plant is capable of full load even without the heat source and therefore is suitable for solar gas turbine plants (see Sections 7.6.5 and 7.6.6).

3.1.4. Gas Turbine Pumped Storage Plants (Compressed Air Energy Storage, CAES)

The Compressed Air Energy Storage (CAES) plant consists of an open-cycle gas turbine and a compressed air storage vessel connected to the gas turbine cycle and located between compressor outlet and furnace chamber. For peak load generation, the compressor power is either partly or fully substituted for by the stored compressed air (Fig. 3.4). The main figure shows an arrangement with sliding air pressure (constant volume, see Section 2.6), while the right-hand figure illustrates an alternative arrangement with pressure kept constant by the geodetic head of a water system [3.3].

The layout and design of CAES plants has some special features such as the following:

- For reasons of storage capacity, the storage pressure is usually chosen well above the optimum upper pressure of an open-cycle gas turbine, the efficiency

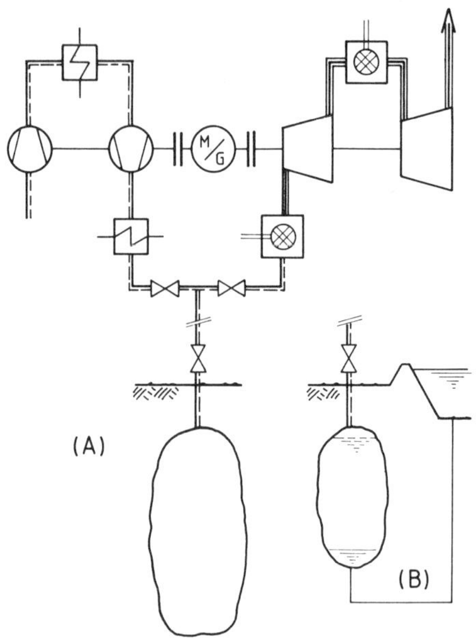

Fig. 3.4. Compressed Air Energy Storage (CAES) plant
A sliding-pressure air accumulator, *B* constant-pressure air accumulator

of the base load process being thus decreased.

- An additional heat store may be arranged between gas turbine and compressed air store where the heat of compression may be stored during the charge operation and regained during discharge. In this way, the storage efficiency is improved and the capacity of the compressed air store is increased.
- A motor or motor/generator is usually provided so as to enable the compressed air store to be charged from the grid (the gas turbine being disconnected) and to permit the operation of the turbine with air from the store only (the compressor being disconnected). In this way, the CAES system becomes a pumped storage system similar to the pure pneumatic storage system to be described in the following section (although it must still burn fuel during discharge).

3.1.5. Pneumatic Pumped Storage

The basic flow sheet is shown in Fig. 3.5. It is similar to the CAES systems. However, there is no furnace chamber.

The separate storage system for the heat of compression is more important for this system than for the CAES. This is so not only for reasons of storage capacity but also in order to avoid condensation or even freezing during the expansion in the turbine. A further heat store may be provided within the air store thus stabilizing the air temperature during compression and expansion in the store itself.

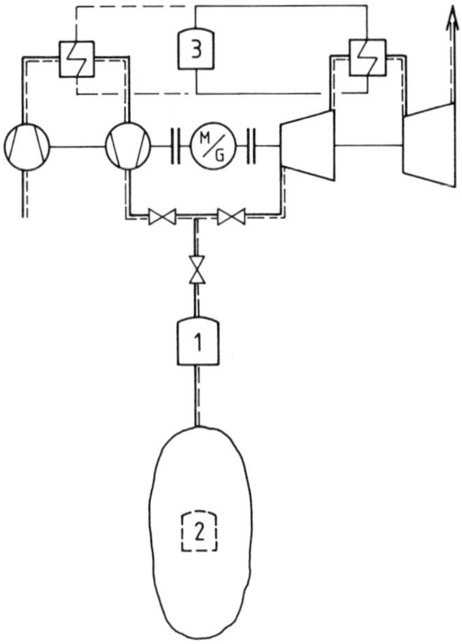

Fig. 3.5. Pneumatic pumped storage plant with separate storage of thermal energy
1 TES in air piping, *2* TES in air storage, *3* TES for intercooling/reheating

3.1.6. Heat Pump Storage Systems

As early as 1933 a heat pump storage system—the so-called thermodynamic storage—was proposed by Marguerre [3.4] for peak load duty, with steam as the working medium and an upper and lower sliding-pressure storage vessel (see Fig. 3.6). For charging, steam is taken from the lower pressure vessel, compressed in the compressor and condensed in the upper storage vessel. During discharge, steam from the upper vessel is expanded in the turbine and condensed in the lower storage vessel. At the charge operation, the pressure in the lower store decreases, while in the upper store it increases.

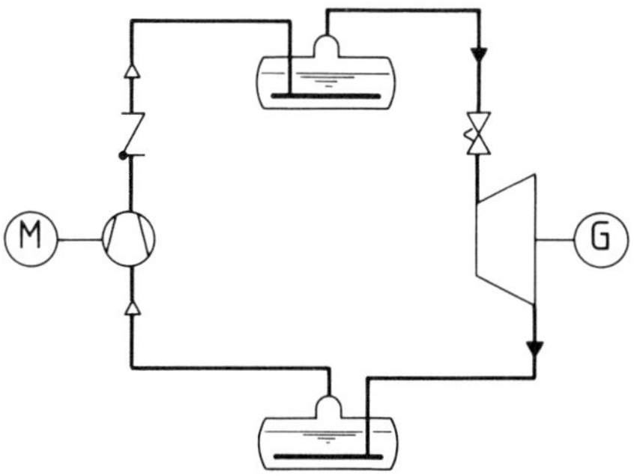

Fig. 3.6. Thermodynamic steam storage [3.4]

Fig. 3.7 shows the basic flow sheet of a heat pump storage system with indirect storage (for instance, a latent heat store) as upper store and the environment as lower store. A refrigerant is used as working medium. For charging, it is evaporated at ambient temperature, compressed in the compressor, and condensed in the upper store. The condensate is throttled down to lower store pressure, thus completing the cycle. For discharging, condensate is pumped through the upper store and evaporated; the steam drives the turbine and is condensed in the lower store.

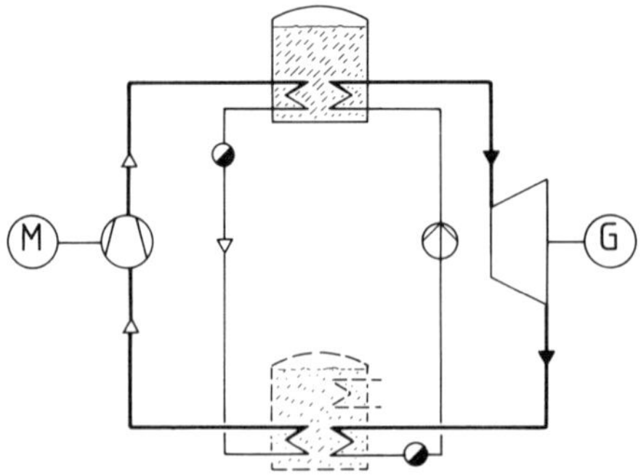

Fig. 3.7. Indirect heat pump storage

3.1.7. Integrated Pumped Storage Systems

Pacault [3.5] has proposed a storage system for Pressurized Water Reactors (PWR) that employs the heat pump principle but is integrated into the plant (Fig. 3.8). A steam compressor (driven from the main shaft) compresses and heats up steam above the temperature level of the PWR. The heat of the condensing steam is transferred by means of hot air to a solid heat store. During discharge, the heat is used to reheat the main steam to a level well above the temperature at base load, thus increasing the useful enthalpy drop in the turbine.

The sorption heat pump system described in Section 2.7 is a further example of an integrated pumped storage system. Mechanical compression is replaced by desorption stemming from increased temperature. Energy storage is direct, the storage fluid also acting as working medium.

3.2. Storage Efficiency

As was mentioned in Section 1.3, the system storage efficiency (turnaround efficiency) is important: High turnaround efficiency means increased peak power, increased utilization and lower fuel consumption. In the following, the turnaround efficiency will be derived from the individual losses and efficiencies for the major types of storage power plants.

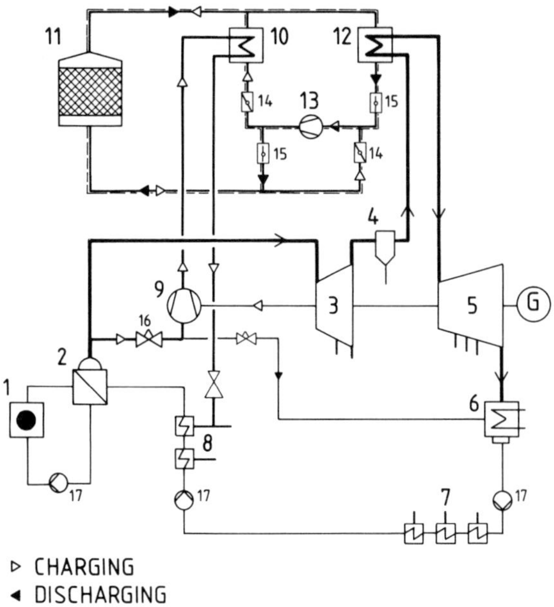

Fig. 3.8. Storage system for pressurized water reactors with steam compressor, according to Pacault [3.5]
1 Pressurized Water Reactor, *2* steam generator, *3* HP turbine, *4* moisture separator, *5* LP turbine, *6* condenser, *7* LP feed-water train, *8* HP feed water train, *9* charging steam compressor, *10* charging steam/air heat exchanger, *11* solid bed accumulator, *12* discharging steam reheater, *13* charging/discharging blower, *14* charging flaps (closed during discharge), *15* discharging flaps (open during discharge), *16* charging steam valve, *17* pumps

3.2.1. Definition

The turnaround efficiency of pumped storage systems η_{st} is defined as the ratio of electric (peak) power supplied during discharge to electric power consumed during charge of a cycle (that is, assuming the same condition in the store at the beginning and at the end). In special cases where mechanical energy is used directly, η_{st} is defined as the ratio of mechanical energies. η_{st} is a ratio of exergies

$$\eta_{st} = E_d / E_c \tag{3.1}$$

η_{st}	turnaround efficiency
E_d	discharge exergy
E_c	charge exergy.

If E_c is taken as the electric energy produced less in the main system due to charging, then the same definition applies to flow storage systems.

The turnaround efficiency consists of three parts

$$\eta_{st} = \zeta_{csys} \zeta_{ac} \zeta_{dsys} \tag{3.2}$$

ζ_{csys} (exergetic) efficiency of charge system
ζ_{ac} (exergetic) efficiency of store, according to Eq. (2.99)
ζ_{dsys} (exergetic) efficiency of discharge system.

The efficiencies of charging and discharging system may be combined to give a system efficiency

$$\zeta_{sys} = \zeta_{csys}\zeta_{dsys} \tag{3.3}$$

which yields

$$\eta_{st} = \zeta_{ac}\zeta_{sys}. \tag{3.4}$$

3.2.2. Pumped Storage Systems

Fig. 3.9 shows a simple pumped storage system. During the charge operation a compressor (pump) increases the exergy of a working medium from ambient ($E = 0$) to E_p. Between the compressor outlet and the store inlet there will be piping losses (moreover, there may be a heat exchanger for indirect heat storage), thereby reducing E_p to E_{in}. The efficiency of the charging system therefore becomes

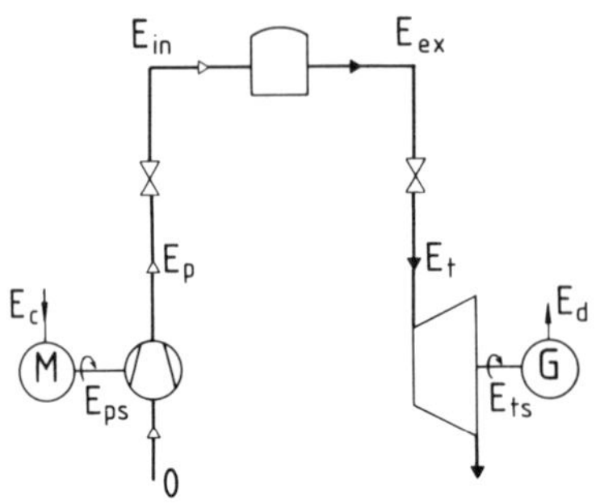

Fig. 3.9. Exergy flow in a pumped storage system

$$\zeta_{csys} = E_{in}/E_c = (E_{in}/E_p)(E_p/E_{ps})(E_{ps}/E_c) = \zeta_C\zeta_P\zeta_M \tag{3.5}$$

in store inlet
c charge
p pump or compressor exit
ps pump or compressor shaft
C charging region
P pump or compressor
M motor.

In a similar manner, the discharge efficiency is found to be

$$\zeta_{dsys} = E_d/E_{ex} = (E_d/E_{ts})(E_{ts}/E_t)(E_t/E_{ex}) = \zeta_D\zeta_T\zeta_G \tag{3.6}$$

ex	store outlet
d	discharge
ts	turbine shaft
t	turbine inlet
D	discharge region
T	turbine (or other expansion engine)
G	generator.

As for the efficiency of the store itself, we have

$$\zeta_{ac} = E_{ex}/E_{in}. \tag{3.7}$$

The turnaround efficiency becomes

$$\eta_{st} = (\zeta_C \zeta_P \zeta_M)\zeta_{ac}(\zeta_D \zeta_T \zeta_G). \tag{3.8}$$

The turnaround efficiency of pumped storage systems thus consists of seven individual efficiencies each of which is less than 1.

3.2.3. Flow Storage Systems

Fig. 3.10 shows a simple flow storage system consisting of a steam generator, a store and an overloadable (peaking) turbine. The following exergy flows $\dot{E}$ take place:

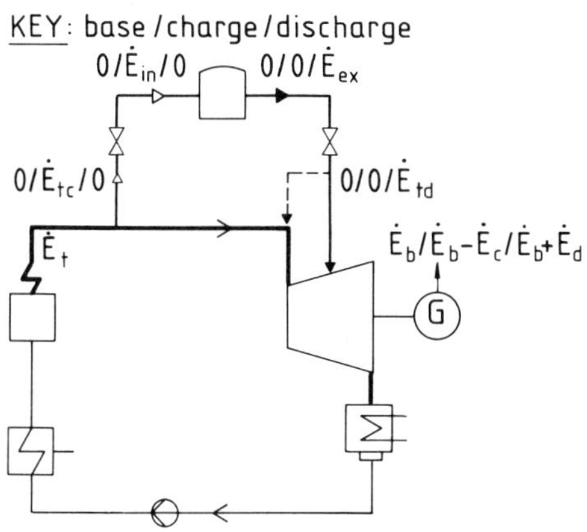

Fig. 3.10. Exergy flow in a flow storage system

(a) Base load:

$$\dot{E}_b = \dot{E}_t \zeta_{TG} \tag{3.9}$$

b	base load
TG	turboset.

(b) Charging:

$$\dot{E}_b - \dot{E}_c = (\dot{E}_t - \dot{E}_{tc})\zeta_{TGc}. \quad (3.10)$$

(c) Discharging via peaking turbine:

$$\dot{E}_d = \dot{E}_{td}\,\zeta_{PT} \quad (3.11)$$

PT peak load turboset.

Since exergy flows $\dot{E}$ are used rather than exergies E (as in the above for pumped storage systems), the integral of $\dot{E}$ over charge and discharge time is to be used for defining turnaround efficiency. If the charge and discharge power is constant, then

$$\eta_{st,\mathrm{flow}} = \dot{E}_d\, t_d / \dot{E}_c\, t_c \quad (3.12)$$

t_d discharge duration
t_c charge duration.

A characteristic number for the charging of flow storage systems is the charge power ratio (that is, the ratio of charge power to base load) given by

$$C = \dot{E}_c / \dot{E}_b \leqslant 1. \quad (3.13)$$

Using the following definitions for the component efficiencies

$$\zeta_{ac} = \dot{E}_{ex}\, t_d / \dot{E}_{in}\, t_c \quad (3.14)$$

$$\zeta_D = \dot{E}_{td} / \dot{E}_{ex} \quad (3.15)$$

$$\zeta_C = \dot{E}_{in} / \dot{E}_{tc} \quad (3.16)$$

and using Eqs. (3.9) to (3.13), the turnaround efficiency of flow storage systems becomes

$$\eta_{st,\,\mathrm{flow}} = \left[\zeta_C\left(1 - \frac{1 - \zeta_{TGc}/\zeta_{TG}}{C}\right) \Big/ \zeta_{TGc}\right] \zeta_{ac}\,[\zeta_D\,\zeta_{PT}]. \quad (3.17)$$

It consists of three part efficiencies and also contains the ratio of peak load turbine efficiency to efficiency of base load turbine during charging ζ_{PT}/ζ_{TGc}.

The term in the parenthesis accounts for the change of turbine efficiencies during charging ζ_{TGc}, compared to the turbine efficiency at normal operation ζ_{TG}, and for C as well. The following extreme cases are possible:

- $\zeta_{TG} = \zeta_{TGc}$: Here, the expression in the parenthesis becomes 1.
- $C \doteq 0$, that is to say, charging occurs during a very long time. According to the rule of de l'Hopital, we have

$$\eta_{st,\,\mathrm{flow}} = \left[\zeta_C\left(1 + \frac{\partial(\zeta_{TGc}/\zeta_{TG})}{\partial C}\right) \Big/ \zeta_{TGc}\right] \zeta_{ac}\,[\zeta_D\,\zeta_{PT}]. \quad (3.18)$$

This equation shows the influence of deviations of turbine efficiency during charging operation.

Fig. 3.11 shows, for an actual case studied [1.25], the dependence of the turnaround efficiency of a sliding-pressure storage in a steam plant when pressure swing—the difference between charge and discharge pressure—and ratio of discharge/charge hours are varied. Assuming constant discharge power and discharge duration, this ratio is proportional to the above-defined charge power ratio C. With increasing abscissa, η_{st} decreases only slowly at the beginning but more strongly later when the turbine efficiency decreases. Large pressure swings lead to large throttling losses during charging and discharging.

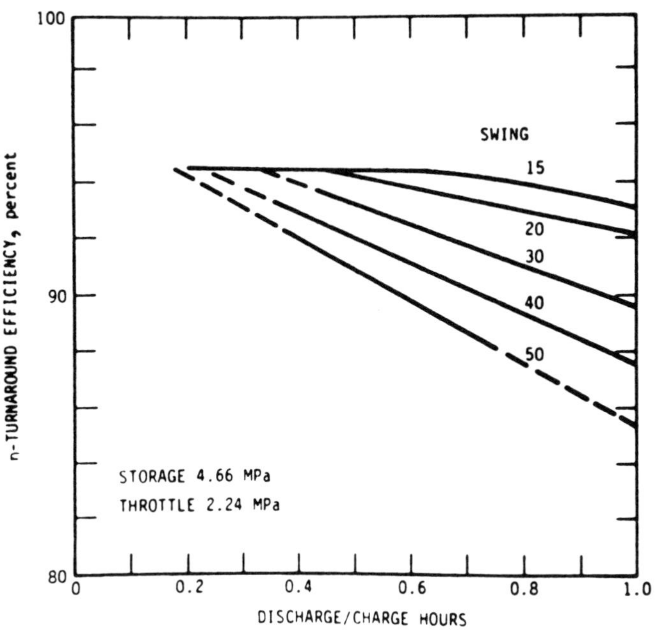

Fig. 3.11. Turnaround efficiency of a sliding-pressure storage plant versus load [1.25]

High turnaround efficiencies are possible with flow storage systems (in which the storage efficiency is composed of only three partial efficiencies and two efficiency ratios). For a constant parameter system, $\dot{E}_{tc} = \dot{E}_{td}$, and $\zeta_C = \zeta_D = 1$. If, moreover, the turboset efficiency at part load and peak load is equal to the turbine efficiency at base load, the turnaround efficiency approaches ζ_{ac}.

3.2.4. Feed Water Storage

Feed water storage is a special case of flow storage and is often applied to steam power plants with regenerative feed water heating. Displacement storage vessels are usually employed, but in principle expansion-type systems also are possible. Fig. 3.12 shows a simplified flow sheet. The storage system is arranged in parallel to the feed water train, taking up feed water during the charging and, at the same time, supplying cold feed water to the train. During discharge, the reverse takes place, reducing bleed flow (in extreme cases, reducing it down to zero) and increasing the power output of the turbine. The following exergy flows appear in Fig. 3.12:

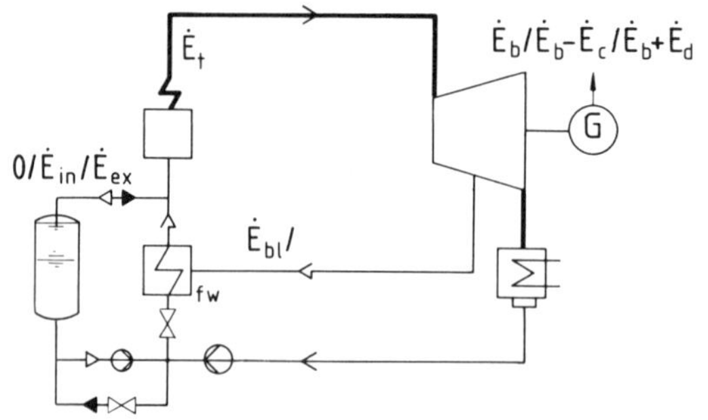

Fig. 3.12. Exergy flow in a feed water storage system

(a) at base load:

$$\dot{E}_b = \zeta_{TG}(\dot{E}_t - \dot{E}_{bl}) \tag{3.19}$$

bl bleed;

(b) at charging:

$$\dot{E}_b - \dot{E}_c = (\dot{E}_t - \dot{E}_{bl} - \dot{E}_{in}/\zeta_{fw,c})\zeta_{TGc} \tag{3.20}$$

where

$$\zeta_{fw} = \dot{E}_{fw}/\dot{E}_{bl} \tag{3.21}$$

fw feed water

is the (exergetic) efficiency of the feed water train;

(c) at discharge (with main turbine):

$$\dot{E}_b + \dot{E}_d = (\dot{E}_t - \dot{E}_{bl} + \dot{E}_{ex}/\zeta_{fw,d})\zeta_{TGd}. \tag{3.22}$$

By arithmetic similar to that in Section 3.2.3 and with the discharge ratio

$$D = \dot{E}_d/\dot{E}_b \leqslant 1 \tag{3.23}$$

the turnaround efficiency becomes

$$\eta_{st,fw} = \left[\zeta_{fw,c}\left(1 - \frac{1 - \zeta_{TGc}/\zeta_{TG}}{C}\right)\Big/\zeta_{TGc}\right] \cdot \zeta_{ac}\left[\zeta_{TGd}\Big/\zeta_{fw,d}\left(1 + \frac{1 - \zeta_{TGd}/\zeta_{TG}}{D}\right)\right]. \tag{3.24}$$

This equation is similar to Eq. (3.17) for flow storage system (see Section 3.2.3), except that now the denominator contains the feed water train efficiency during discharge and a further factor (in parenthesis) contains the efficiency change of the turboset during discharge. It should be noted that the discharge system efficiency

$$\zeta_{dsys} = \zeta_{TGd} \Bigg/ \zeta_{fw,d} \left(1 + \frac{1 - \zeta_{TGd}/\zeta_{TG}}{D}\right) \tag{3.25}$$

may be larger than 1 (for example, if feed water train efficiency $\zeta_{fw,d}$ is low). On the other hand, a low efficiency of the feed water train lowers the efficiency of the charge system ζ_{csys}, affecting the turnaround efficiency, and also the thermal efficiency of the main process.

Only in simple cases such as the ones above may the turnaround efficiency be assessed from partial efficiencies. More complex systems will require the use of computer programs for the thermodynamic cycles.

3.2.5. Losses in Storage Power Plants

Table 3.1 specifies the types of losses in the more important types of energy storage power plants. The plant is characterized by the type of storage as well as by the storage system (direct or indirect, flow or pumped storage). Losses of the store itself and of the charge and discharge equipment are specified. Also the discharge efficiency is given which is important for determining—together with the exergy density (see Section 2.8.2)—the discharge exergy and therefore the size of the store. The turnaround efficiency of the whole plant is also given.

Direct storage of the working medium or the heat transfer medium—particularly feed water storage in steam plants and primary cycle storage in solar power plants—yields rather high values of turnaround efficiency. The high value of 0.95 for sliding-pressure storage applies only to small pressure swings that require inexpensive pressure vessels.

Turnaround efficiencies of indirect storage systems are lower. Storage of sensible heat may be superior to latent heat storage where heat transfer from the working medium to storage medium and back requires costly equipment and entails exergy losses. The constant storage temperature is an advantage only for the steam process; it is a disadvantage for the gas turbine process. Thermochemical storage has rather low efficiencies but the advantage is that there is no insulation loss and therefore no deterioration during prolonged periods (seasonal storage). A direct thermochemical storage system could be interesting.

The turnaround efficiency of pumped storage systems is low because of the seven part efficiencies, the hydraulic ones being superior to the pneumatic ones. Constant pressure pneumatic storage systems avoid throttling losses during charging and discharging. The turnaround efficiency of heat pump storage systems with sliding-pressure accumulators suffers from the pressure swings in the upper and lower storage vessel and the corresponding compressor and turbine losses.

Table 3.1. *Losses and Efficiencies of Storage Power Plants*

Storage Power Plants			Exergy Losses			Exergetic Efficiency of Discharge	Storage Efficiency (turnaround)
Type of Storage		Storage System	Storage Losses	System Losses			
				Charging	Discharging		
Saturated liquids	Sliding pressure	Direct flow storage with flash evaporation (stepwise)	Static head, insulation, stratification	Pressure drop, piping, throttling	Throttling, turboset efficiency difference, condenser pressure increase	0.7–0.97	0.5–0.95
		Direct heat pump storage	Charging by super-heated steam, static head of upper store, insulation, charging losses of lower store	Motor, compressor, pressure drop, throttling	Throttling, pressure drop, turbine, generator	0.7	0.45–0.5
	Expansion storage	Direct flow storage with flash evaporation (stepwise)	Insulation	Heat exchange or mixing, pumping	Heat exchange or throttling	0.8–0.95	0.65–0.9
Subcooled liquids	Hot water displacement		Insulation, mixing, temperature equalization		Heat exchange or throttling, pumping, turboset efficiency difference, condenser pressure increase	0.8–0.95	0.7–0.9
		Direct feed water storage		Incomplete preheating, pumping	Pumping, turboset efficiency difference, condenser pressure increase	0.9–1.02	0.8–0.95
	High boiling liquids (one vessel; multi-vessel)	Indirect feed water storage	Insulation, heating and cooling losses of multi-vessel systems	Heat exchange, pumping	Heat exchange, pumping, turboset efficiency difference, condenser pressure increase	0.75–0.85	0.6–0.75
		Direct flow storage with primary cycle heat storage		Pumping	Heat exchange, turboset efficiency difference, condenser pressure increase	0.9–0.97	0.8–0.95

Solids		Indirect flow storage	Insulation, temperature equalization	Pressure drop, heat exchange, pumping	Pressure drop, heat exchange, pumping, turboset efficiency difference	0.8–0.85	0.7–0.75
Latent heat			Insulation, subcooling			0.75–0.8	0.6–0.65
Thermochemical			Charging/discharging hysteresis	Pressure drop, heat exchange, pumping, cooling down, condensing or compressing of reaction partners	Pressure drop, heat exchange, pumping, heating up, evaporation or throttling of reaction partners	0.7–0.74	0.5–0.55
Geodetic head of water		Hydraulic pumped storage	Leakage, evaporation	Motor, pumping, pressure drop	Pressure drop, turbine, generator	0.7–0.8	0.5–0.75
Compressed air	Sliding pressure	Pneumatic pumped storage	Cooling down, leakage, condensation of moisture	Motor, compressor, pressure drop, throttling, cooling down	Throttling, pressure drop, turbine, generator	0.82	0.65
	Constant pressure			Motor, compressor, pressure drop, cooling down	Pressure drop, turbine, generator	0.84	0.7
	Sliding pressure	Gas turbine pumped storage (CAES)		Motor, compressor, pressure drop, throttling, cooling down	Throttling, pressure drop, turbine, combustion loss and stack loss difference, generator	–	0.55[1]
	Constant pressure			Motor, compressor, pressure drop, cooling down	Pressure drop, turbine, combustion loss and stack loss difference, generator	–	0.58[1]

[1] See Section 7.3.2.

It follows from Table 3.1 that in general the turnaround efficiencies of the more conventional thermal storage systems are better than those of the new technologies such as latent heat storage, thermochemical, pneumatic and heat pump storage. The difference can be offset only by much lower installation costs of the new systems.

3.3. Limitations of Charge and Discharge Power

As was pointed out in Section 1.3, optimum operation of a storage power plant depends not only on its storage capacity but also on the discharge and charge power and their possible rate of change. Limitations of power and its rate of change may, to some extent, be a consequence of the type of the storage plant.

3.3.1. Limitations by Plant Layout

The limitation of charge or discharge power by the layout of charge and discharge equipment is trivial. Layout of equipment may be the result of an optimization in the grid and may be different for different storage systems: The charge equipment of a pneumatic storage system may, for reasons of installation cost, be designed for a power much smaller than discharge power (even if the system thus looses some of its flexibility). On the other hand, flow storage systems with direct storage will be designed for high charge powers, since this does not greatly affect investment cost.

3.3.2. Limits of Components

Limitations may also be imposed by limits on the size of components such as single-shaft turbines (presently 300 MW for gas turbines without compressor and 1200 MW for steam turbines). This applies not only to the CAES plant but also to steam plants with integrated storage and overloadable main turbine. The limits may be either technological or due to considerable efficiency decreases at extreme loads. However, an advantage of the overloadable main turbine is its high flexibility (that is, the possibility of rapid load changes, in contrast to the separate peaking turbine which requires spinning in the hot condition in order to be able to pick up load quickly).

Further limitations may be imposed by limits on the size of the pressure vessel, the pumps and so on. These, however, may be overcome by multiple units.

3.3.3. Thermodynamic Limitations

These apply to integrated heat pump storage plants and to feed water storage.

In the integrated heat pump storage plant [3.5], the steam flow rate is kept constant and peak load is achieved by an increase in steam temperature only. Therefore, the rate of load change is limited by the allowable rate of temperature change of the turbine (in the proposal of Pacault, described in Section 3.1.7, the discharge ratio is 25%). There is no limit for the charge power except that of the investment cost of the steam compressor (in the example: 15% compressor power from shaft plus loss of turbine power because of compressor steam consumption, resulting in 32% relative charge power).

The thermodynamic limitation of the discharge power of feed water storage is given by the exergy of the bleed steam flows used for feed water heating during normal (base load) operation. Steam temperatures are more or less constant, although steam flow rates are variable. Discharge power is limited by the following parameters:

(1) Feed water temperature: With pure feed water storage, this parameter is equal to the upper storage temperature. The higher it is, the higher is the relative discharge power. Feed water temperature is, however, sometimes limited to values below the thermodynamic optimum (for instance, in Light Water Reactors on account of core heat transfer).

(2) Maximum portion of feed water flow to be replaced by stored feed water: In saturated steam turbines, the bleed flow also takes care of moisture reduction and therefore cannot be reduced to zero. Furthermore, the condensate of live-steam heated reheaters–as in Light Water Reactors–cannot be replaced by stored hot water.

(3) Lower storage temperature: Sometimes the lower storage temperature is, for operational reasons (for example, degassing), chosen as the temperature of the lowest or lowest but one bleed stage and is thus higher than the condensation temperature.

(4) Efficiency of the feed water train: As was pointed out in Section 3.2.4, relative discharge power is increased by low feed water train efficiency.

(5) State of live and reheat steam: With increased steam pressure and temperature (and therefore increased enthalpy drop in the turbine), the relative discharge power decreases (for fixed feed water temperature).

Table 3.2. *Typical Data of Steam Cycle* [1.29, 7.25]

Type of Plant	Live Steam		Reheat Steam		Feed Water Temperature	Ratio of Feed Flow Replacement to Live Steam Flow Rate
	Pressure bar	Temperature °C	Pressure bar	Temperature °C	°C	
CANDU	55	sat.	10	250	180	0.82
LWR	70	sat.	12	265	210	0.79
HTR	170	540	40	540	206	1.0
LMFBR	170	540	40	540	240	1.0
Fossil, subcritical	170	540	40	540	240	1.0
Fossil, supercritical	250	540	55	540	260	1.0

The main data of the power plants considered are given in Table 3.2, while Table 3.3 examines the relative discharge power D.

Table 3.3. *Relative Discharge Power of Feed Water and Cascading Storage* [1.29, 7.25]

Type of Plant	Relative Discharge Power $D = \Delta N/N$ (% of Base Load)					
	Feed Water Storage	Cascading Storage, at an Upper Pressure of (bar)				
		20	55	70	100	170
CANDU	7.6	12.4	23.9	–	–	–
LWR	13.0	13.3	26.0	30.2		
HTR	8.6	9.3	17.1	20.0	26.7	42.0
LMFBR	13.4	–	18.0	21.1	27.1	44.0
Fossil, subcritical	13.4	–	18.0	21.1	17.1	44.0
Fossil, supercritical	17.2	–	19.0	22.2	28.2	47.0

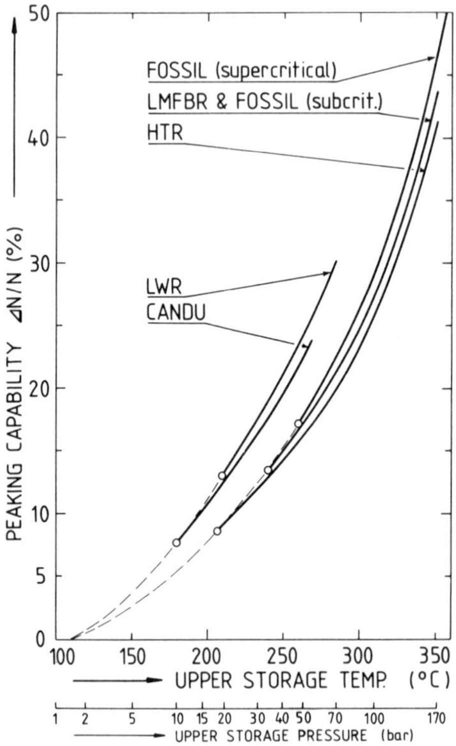

Fig. 3.13. Relative discharge power $D = \Delta N/N$ of pure feed water and cascading storage

With storage temperatures above the feed water temperature, the relative discharge power may be increased above the one which applies to pure feed water storage by means of cascading storage (see Section 3.1.2). This is shown in Fig. 3.13, where a lower storage temperature of 110°C is assumed. The limits for pure feed water storage are: about 8% for Heavy Water Reactor (CANDU) and High Temperature Reactor (HTR); about 13% for the Light Water Reactor (LWR), for the Liquid Metal-cooled Fast Breeder Reactor (LMFBR) and for subcritical fossil-fired plants; about 17% for supercritical fossil plants. With cascading storage at about 70 bar, these limits may be extended considerably [3.2].

There is also an ultimate thermodynamic limitation on the charge power, since the sum of the bleed flows cannot be increased beyond the live steam flow rate.

3.4. Control and Load Distribution in Storage Power Plants

As in any power plant, the main task of control of the storage plant is to supply a predetermined power (load) to the grid or to the consumer. This load may be defined by external load management or by an (internal) load controller using signals from the grid, the generator speed, the set load and the actual load.

There are two special features to be considered:

- In a storage power plant, not only (positive) discharge power but also (negative) charge power is to be controlled.
- The charge condition of the store is an important parameter for optimum operation. It is either to be made known to the external load manager or it is to be integrated into the control scheme in case of internal load control.

The control concept itself will depend on operational conditions and on the type of storage system [3.6].

3.4.1. Non-integrated Storage Plants

By means of the turbine inlet valve, load control of a non-integrated storage plant with separate energy converter and separate turbine is achieved in a manner that is both simple and similar to constant pressure control in a steam power plant. Fig. 3.14 explains the principle.

The controller governing the turbine valve may use either a load signal from a comparison of set and actual loads or a grid signal (for instance the frequency). A limitation of load change rate will not normally be required. In case of sliding-pressure storage systems, the pressure decrease during discharge will be a perturbation which can be eliminated by an interference factor compensation (dashed line) or by an additional throttling valve, which is pressure-controlled (dotted line).

Fig. 3.15 shows the load control system of a flow storage plant with separate peak load turbine in some detail. There is a displacement storage vessel, which is

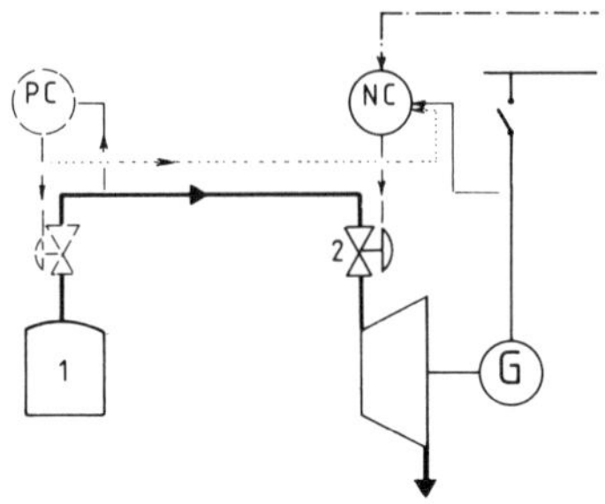

Fig. 3.14. Load control of a non-integrated storage plant
NC load controller, *PC* pressure controller, *1* store, *2* turbine valve

supplied with hot water from a mixing preheater which in turn gets steam from the steam generator. For discharge, a train of flash evaporators is provided, similar to (A1) in Fig. 3.2. The load signal acts on the turbine valve *V*5. The storage system being at constant pressure and the throttle valve *V*6 (in front of the high-pressure flash evaporator) producing constant pressure, no interference factor compensation is now necessary. Further throttle valves *V*7, *V*8 and *V*9 control the water level in the flash evaporators. This level control may be a three-element control (as for a recirculation boiler) acting on the mass flow rates of the flash evaporator and their difference. The last flash evaporator acts as de-gasser and has a temperature control and a level control [4.11].

In pneumatic pumped storage plants, a control of charge power is possible but is restricted by the compressor characteristic.

3.4.2. Integrated Storage Plants

The load demand in integrated storage plants is covered jointly by the base load part and the storage (peak load) part of the plant. The load distribution between these two parts is important.

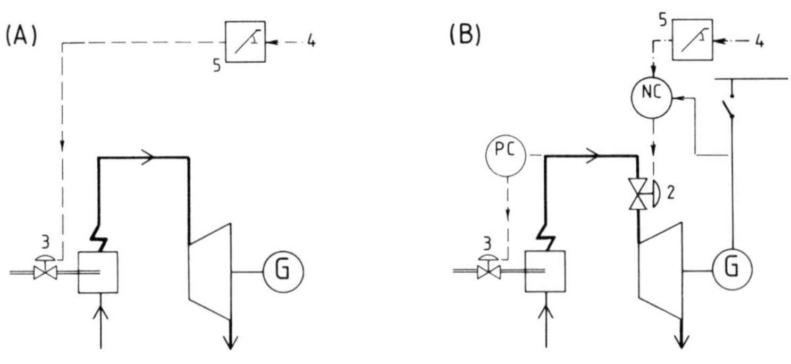

Fig. 3.16. Load control of a steam plant without storage
A open loop control, *B* load control with feedback and constant boiler pressure operation, *2* turbine valve, *3* fuel valve, *4* power demand signal, *5* signal limiter

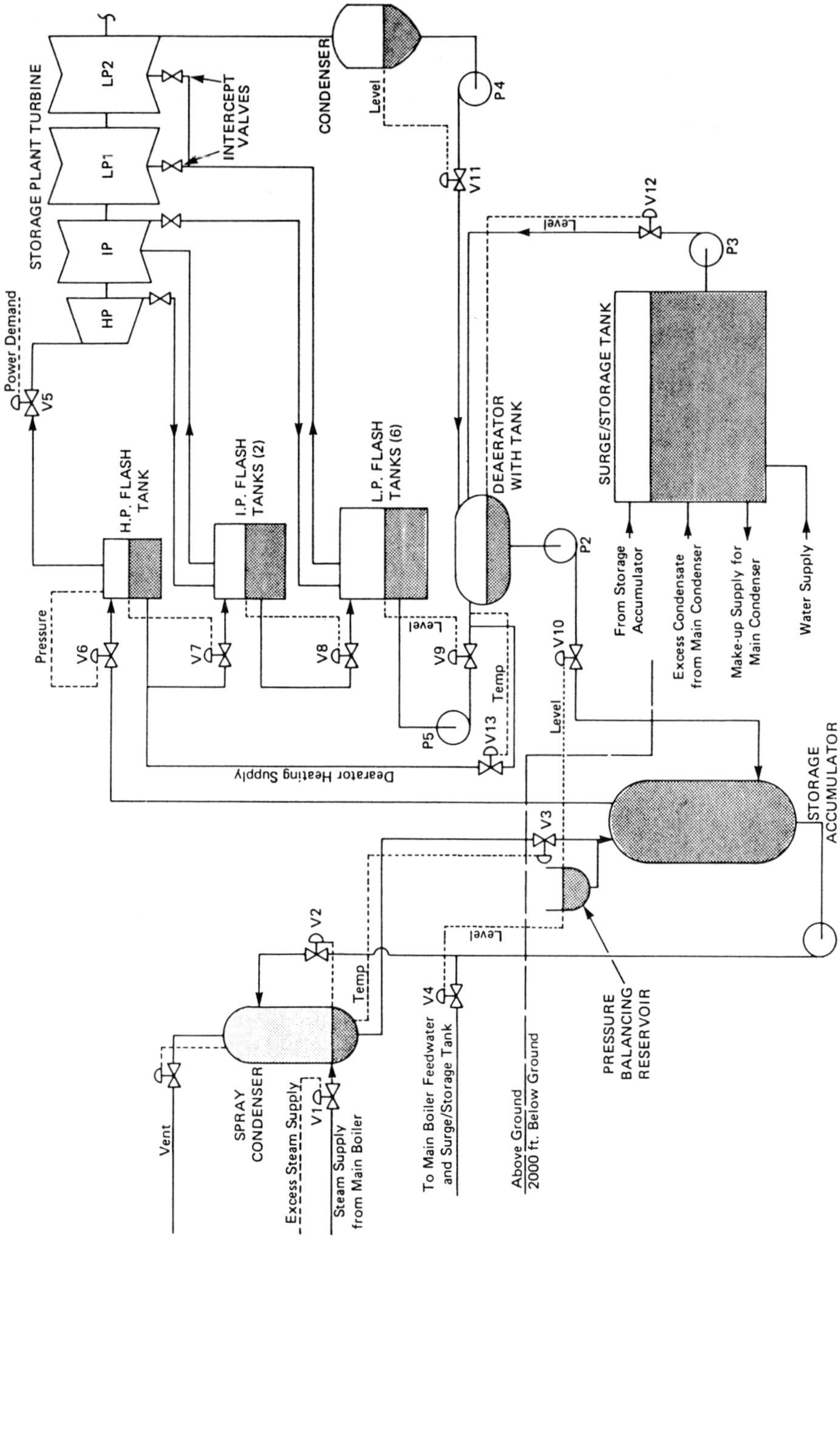

Fig. 3.15. Control system of a displacement storage plant [4.11]

The two basic possibilities of load control of a steam plant *without* storage are shown in Fig. 3.16:

(a) The load controller acts on the fuel supply, while the turbine uses the supplied steam more or less without control (see Fig. 3.16 A). Considering the inertia of the system, the speed of load change should be limited.

(b) The load controller acts on the turbine valve, while boiler control ensures constant live steam pressure (see Fig. 3.16 B). The rate of load change should be limited in order to limit the temperature gradients stemming from the pressure gradients in the turbine.

Load control of power plant *with* integrated storage exhibits the following special features:

(1) If the fast changes are taken up by the storage part, a load change limitation may not be necessary.

(2) Boiler control gets the charge condition of the storage system as an additional signal, since the boiler must not only supply steam to the turbine but to the storage part also.

(3) The control system must distribute the total load between base load part and storage (peak load) part according to their characteristics.

The scheme in Fig. 3.17 shows the load distribution between boiler and store in an integrated storage steam power plant. The load signal from the load

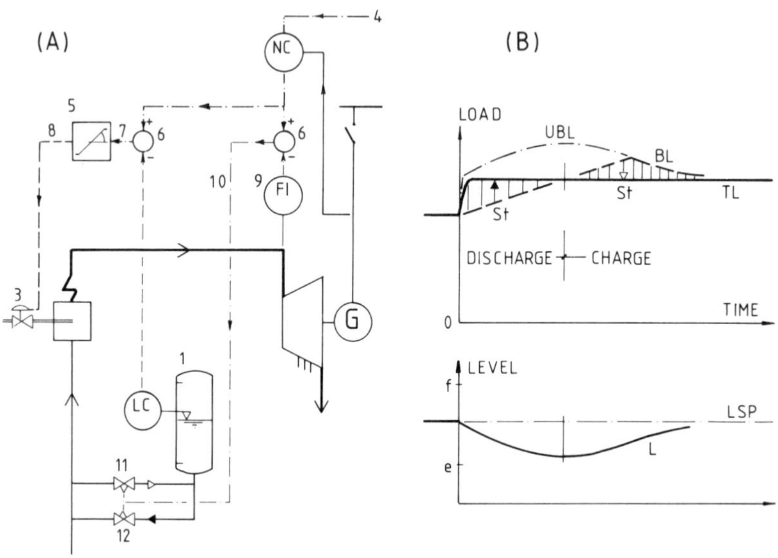

Fig. 3.17. Load distribution between boiler and store
A control system: *1* store, *2* turbine valve, *3* fuel valve, *4* power demand signal, *5* signal limiter, *6* difference amplifier, *7* unlimited boiler load signal, *8* boiler load signal (limited), *9* actual boiler load, *10* store load signal, *11* charge valve, *12* discharge valve, *LC* level controller of store; *B* load responses on a step input: *TL* turbine load, *UBL* unlimited boiler load, *BL* boiler load, *St* load of store, *L* level of store, *f* full, *e* empty, *LSP* level set point

controller is compared to the instantaneous actual boiler load. The difference generates a load signal to the charge or discharge equipment of the store and its task is to cover this difference (Fig. 3.17 A). The load signal and a signal for the charge condition of the store are combined to produce a signal for the boiler load.

This boiler load signal will get a load limitation (so as to avoid boiler overload or uneconomic operation at extreme part loads) as well as a limitation of load change rate. A load distribution control and a simple boiler load limitation has been shown in Fig. 1.4. As concerns a sudden change of load demand and a limitation of boiler load change rate, Fig. 3.17 B shows the reactions of load of storage plant, boiler load and charge condition. It turns out that even a small storage capacity may, with power control, improve the load behaviour markedly. For the base load part, a simple control concept should be chosen, perhaps even without a turbine control valve (if the storage plant takes care of all the fast load changes completely).

There are other possibilities for the load distribution between base load part and storage part: For instance, the rate of load change may be measured and may give to the storage part an advance load signal that is made to correct its actual load, thereby avoiding any major control signal to the slow control of the base load plant [3.6]. However, care must be taken to avoid the situation in which the storage plant load exceeds the demand load change, for in this situation the base load plant would receive a signal in the wrong direction.

3.4.3. Steam Power Plant with Buffer Storage

The task of buffer storage is to allow fast load changes of the turbine without a load change of the boiler and its furnace. In order to cater to load increases and decreases, the charge condition should be medium in normal operation. Buffer storage is short-term storage, the more prolonged load changes being met by boiler load control.

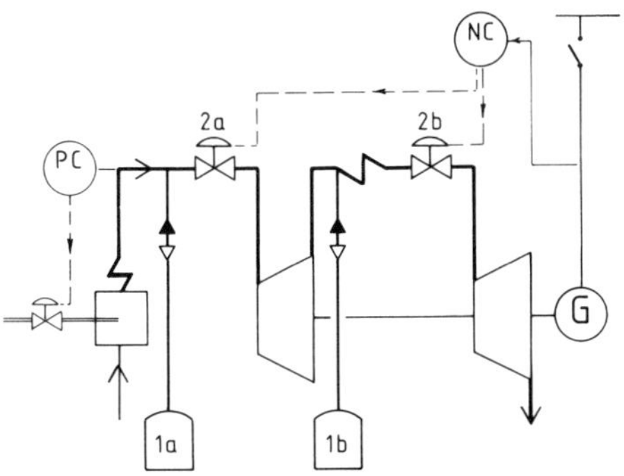

Fig. 3.18. Load control of a steam power plant with buffer storage
1a HP buffer store, *1b* IP buffer store, *2a* HP turbine valve, *2b* IP turbine valve

Fig. 3.18 shows a flow sheet with sliding-pressure storage connected to live steam and/or cold reheat steam. The load controller acts on the turbine valve in front of the HP and LP casings. Boiler control maintains live steam pressure and, therefore, the charge condition of the sliding-pressure storage vessel. Unless high throttling losses in the turbine valve are accepted, the useful pressure changes are rather low.

Operation in case of a sudden demand increase is as follows: The turbine valves open, the steam pressure in the piping to the turbine valve decreases and the storage vessels discharge steam, thus taking over the load increase for a short time. Finally, the boiler control system increases the boiler load, the boiler load controller limiting the load and the rate of load change.

4. Storage Vessels

The storage vessel is an important part of the thermal energy storage system. In particular, if it is pressurized, then national laws, codes and regulations will have to be obeyed. The storage vessel has several functions [4.1]:

- to seal the storage medium against ambient;
- to permit mass and/or heat exchange for charging and discharging operations;
- to resist the pressure of the storage medium, including the hydrostatic pressure or (in the case of indirect storage) the pressure of the heat transfer medium, and to ensure dimensional stability;
- to minimize thermal losses to ambient by proper external or internal thermal insulation;
- to resist corrosion stemming from possible chemical attack of the storage medium.

The manner in which these functions are fulfilled defines the various types of storage vessels:

Welded cylindrical or spherical pressure vessels are of relatively simple design, having only a few penetrations. The vessel wall has the functions of sealing and resisting pressure including dimensional stability. For corrosion protection, either an internal protective cover or—for slow and constant corrosion rates—an increase of wall thickness is provided. Thermal insulation at high temperatures is located internally (thus subjecting the vessel only to pressure-induced loads) or at low and medium temperatures it is usually external (thus subjecting the vessel to loads due to pressure and temperature).

Underground pressure vessels are designed in such a way that the surroundings (subsoil, rock, salt, etc.) will take at least part of the load. It has recently been proposed that under favourable geological conditions such underground caverns of very large size be built. For pressurized air storage, a few have actually been constructed.

Prestressed pressure vessels have recently been proposed for above-ground energy storage at very high pressures and/or in very large sizes. Load is taken by high strength tendons which are prestressed to such an extent that the vessel wall is under compression. This makes possible the use of concrete or cast iron vessel walls. A high-pressure helium energy storage vessel of the Prestressed Cast Iron Vessel (PCIV) type has already been built.

The primary load for "*unpressurized*" vessels, that is for storage media that do not require pressures above atmospheric, will be hydrostatic pressure. Moreover, there may be secondary stress because of different thermal expansion, especially in the case of Thermocline storage. There is a whole spectrum of unpressurized vessels, from steel tanks and sealed concrete tanks to hot water lakes and caverns or aquifers.

4.1. Vessel Shape

The shape of pressure vessels usually is symmetric by rotation. The vessels are mostly cylindrical, with dished heads, in a horizontal or vertical arrangement; sometimes they are spherical. Unpressurized vessels are usually cylindrical with flat lower head and slightly conical upper head (tank shape).

An important characteristic of cylindrical vessels is the slimness

$$S = H/D \tag{4.1}$$

H internal *height* of vertical; internal *length* of horizontal vessel
D internal diameter.

The shape, slimness and volume of the vessel will determine its investment and operating cost [4.2, 4.3]. Specific cost will generally decrease with increasing vessel volume but there will be technical and economical limits (for example, as regards wall thickness, diameter, height and weight).

4.1.1. Optimum Material Utilization

A "dimensionless material requirement" V_B^* which depends on S but not on volume, pressure, or strength may be defined for pressure vessels whose internal pressure is large in comparison to the hydrostatic pressure, the vessel wall taking the load due to internal pressure. This requirement is given by

$$V_B^* = \frac{V_B \sigma}{V_i p_i} \tag{4.2}$$

where

V_B	material volume of wall
σ	allowable stress (N/mm^2 = MPa)
V_i	internal vessel volume
p_i	design pressure (N/mm^2 = MPa)
$V_i p_i$	"pressure volume" of vessel
$V_B \sigma$	"stress volume" of vessel wall.

For the cylindrical vessel having two hemispherical heads:

$$V_i = (D^2 \pi/4)(H - D) + (4/3)\pi(D/2)^3 = (D^3 \pi/4)(S - 1/3) \tag{4.3}$$

$$O_M = D\pi(H - D) = D^2 \pi(S - 1) \tag{4.4}$$

$$O_H = D^2 \pi \tag{4.5}$$

$$s_M = \frac{D_a p_i}{2(K/\nu)\upsilon + p_i} + C_1 + C_2, \qquad s_H = \frac{\beta D_a p_i}{4(K/\nu)\upsilon + p_i} + C_1 + C_2 \tag{4.6}$$

where

O	surface area
s	wall thickness (according to the German Code [4.4])
D_a	outer diameter
p_i	design pressure (N/mm^2 = MPa)
β	shape factor ($\beta = 1$ for hemispherical head)
K	yield stress (N/mm^2 = MPa)
ν	safety factor (> 1)
υ	welding factor (< 1)
C_1	additional wall thickness for minus tolerance
C_2	additional wall thickness for corrosion.

Indices:

M	cylinder
H	the two heads.

Assuming the following simplifications

$$C_1 = C_2 = 0$$

$$K \gg p \qquad \text{and} \qquad s \ll D_a$$

we obtain

$$\sigma = (K/\nu)\upsilon \tag{4.7}$$

$$s_M = Dp_i/2\sigma \tag{4.8}$$

$$s_H = \beta(Dp_i/4\sigma) \tag{4.9}$$

$$V_B = O_M s_M + O_H s_H \tag{4.10}$$

and the dimensionless material requirement V_B^* for cylindrical vessels with hemispherical heads becomes

$$V_B^* = (2S - 2 + \beta)/(S - 1/3). \tag{4.11}$$

In Fig. 4.1 it is shown that $S = 1$ is the sphere with $V_B^* = 1.5$. Values of V_B^* for hemispherical ends with (a) half the cylinder wall thickness [$\beta = 1$] and (b) full cylinder wall thickness [$\beta = 2$] are shown. Dished ends flatter than hemispheres require material according to (b), that is, $\beta = 2$. In this case, high values of S are an advantage, whereas for $\beta = 1$ they are a disadvantage. In any case, V_B^* approaches 2.0 for very slim vessel ($S \to \infty$). For large vertical vessels, the hydrostatic pressure may be important, since it shifts the optimum to lower values of S.

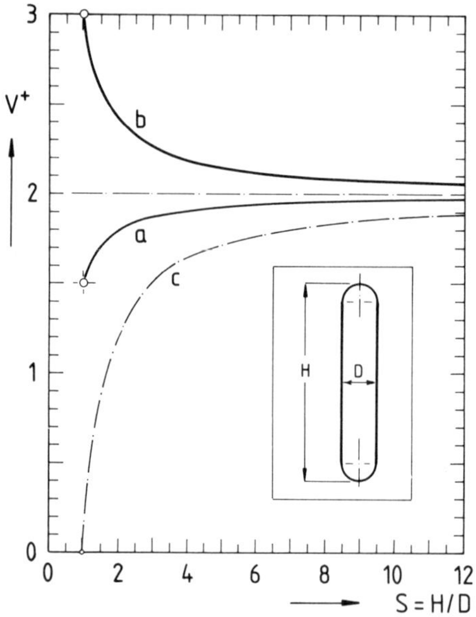

Fig. 4.1. Dimensionless material requirement $V_B^* = V_B \sigma / V_i p_i$ of cylindrical vessels with hemispheric heads versus slimness $S = H/D$, according to Eq. (4.11)
a heads with half wall thickness of shell, *b* heads with full wall thickness of shell, *c* shell only

4.1.2. Optimum Surface Area

The vessel surface area has an effect on first cost via the additional wall thickness, a possible liner or cover for protection against corrosion, and the thermal insulation. It also affects running cost via the heat loss. There is, therefore, an incentive to keep the surface area low. A "dimensionless surface area" O^* may be defined as follows:

$$O^* = O/V_i^{2/3} \tag{4.12}$$

$$O = O_M + O_H . \tag{4.13}$$

This is plotted in Fig. 4.2 in (a) for hemispherical and in (c) for flat heads versus S, where we again assume that $s \ll D_a$ ("thin" walls). Again, the sphere is best. However, in contrast to the dimensionless materials utilization V_B^*, the specific surface area does not approach an asymptotic value but rather increases with S in a monotone way and, moreover, it depends on the absolute volume of the single vessel

$$o = \frac{O}{V_i} = O^*/V_i^{1/3} . \tag{4.14}$$

From the viewpoint of minimizing the surface area, one large single vessel is therefore best.

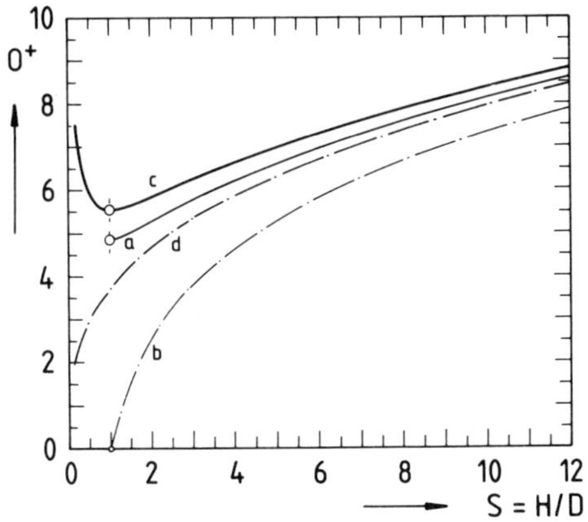

Fig. 4.2. Dimensionless surface area O^* of cylindrical vessels versus slimness $S = H/D$
a with hemispheric heads, *b* shell portion of (a), *c* with flat heads, *d* shell portion of (c)

4.1.3. Limitations

There are, of course, technical and economical size limitations, for instance:

- There may be a limitation on wall thickness in that above a certain wall thickness s_{max} (given in the code) annealing of the walls or of the whole vessel may be required. This would raise the cost of vessels considerably, particularly if some of the welds are done on site. For a given type of steel, this limitation amounts to a limitation on the vessel diameter given by

$$D \leqslant D_{max} = \frac{2 s_{max} \sigma}{p_i} . \tag{4.15}$$

- Diameter limitations may also exist for reasons of fabrication or transportation.
- For "unpressurized" vessels, a limitation on D and H is given by the hydrostatic pressure of the storage medium with density ρ:

$$p = \rho g H \tag{4.16}$$

$$DH \leqslant (DH)_{max} = \frac{2 s_{max} \sigma}{\rho g} . \tag{4.17}$$

- Limitations may also exist for the absolute height H (because of wind pressure, air traffic, etc.) or for the foundation pressure which, according to Eq. (4.16), is also a function of H.

4.1.4. General Optimization

The actual optimum of D and H with regard to the surface area will now depend on the limitations applicable to the individual case. Fig. 4.3 shows a

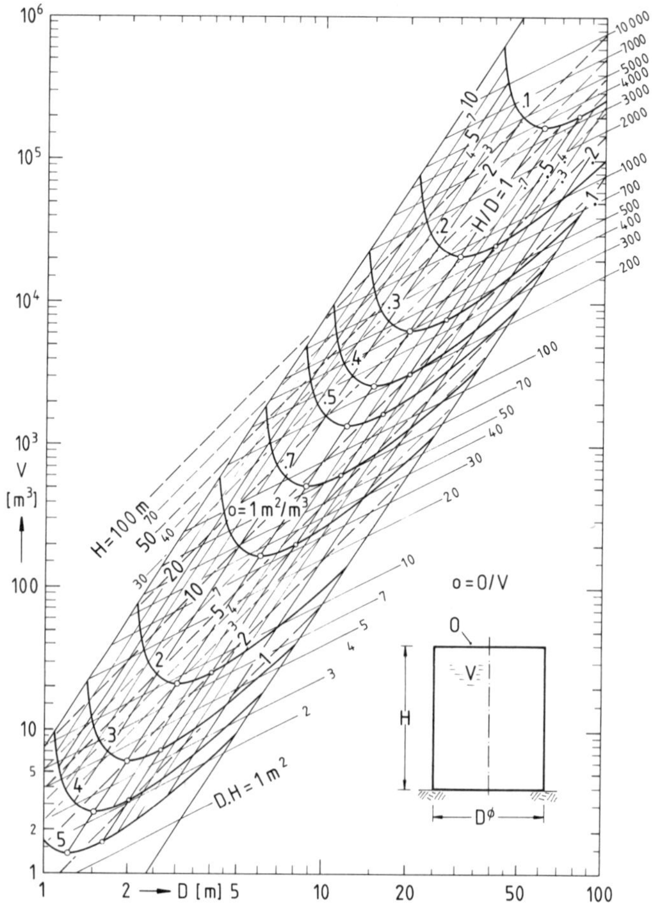

Fig. 4.3. Volume V of cylindrical vessels with flat heads versus diameter D

double logarithmic plot of V versus D, with H as parameter, for cylindrical vessel with flat heads. DH = constant, H = constant, $H/D = S$ = constant are bundles of parallel lines. Moreover, curves of the families o = constant are plotted, where o depends on D, H and S as follows:

$$o = \frac{O}{V_i} = \frac{D\pi H + 2D^2\pi/4}{D^2\pi H/4} = \frac{4}{D} + \frac{2}{H} = \frac{1}{D}(4 + 2/S). \tag{4.18}$$

Because of the double logarithmic plot, the lines of o = constant are congruent and are arranged in a parallel fashion along the line $H/D = S$ = constant.

The vessel shape and size having minimum surface area is obtained by introducing the limitation(s) into the diagram and searching for $o_{\min}$ along the lines of limitation. The result may be a real minimum given by the tangent on the o-curve, or it may be a border-line minimum. In particular:

– For a vessel with given volume, $o_{\min}$ is found at the tangent of the line

$V = V_1$ = constant with the o-line, that is, at $S = H/D = 1$.

- For an unpressurized tank, o_{min} is found where the line of $(DH)_{max}$ touches the o-line. This is always the case at $S = H/D = 0.5$. It may also be found by introducing Eq. (4.17) into Eq. (4.18) and letting the differential become zero. This yields o_{min} at

$$D = \sqrt{2}\sqrt{\frac{2s_{max}\,\sigma}{\rho g}}, \qquad H = \frac{1}{\sqrt{2}}\sqrt{\frac{2s_{max}\,\sigma}{\rho g}}. \tag{4.19}$$

  Since s_{max} is utilized, this case of o_{min} is, at the same time, the case of minimum V_B^*.
- For a vessel with limited diameter and unlimited volume, $S = \infty$ yields o_{min} as a border-line minimum. This is, of course, not quite correct for vertical vessels where hydrostatic pressure should be accounted for.
- A limitation of H leads, at unlimited V, to a border-line minimum at $S = 0$, provided there is no other limitation.

Preselecting size and shape of the vessel(s) according to minimum requirement of material (that is, minimum V_B^*, using Fig. 4.1) on the one hand, and according to minimum surface area o_{min} (using Figs. 4.2 and 4.3) on the other will lead to somewhat contradictory results in many cases. A final optimization will have to consist of a specific parametric study or at least a comparative cost analysis of some promising alternatives which considers the actual differential investment cost of the vessel itself and of the liner (if any), corrosion protection, thermal insulation, support, etc. as well as the differential cost of thermal losses or energy losses as required.

Vertical pressure vessels with dished ends are generally designed with $S \doteq 4$, while horizontal ones are designed with $S > 4$, particularly if there is a diameter limitation. Unpressurized vessels with flat heads are designed with $S \doteq 1$ (if there is no wall thickness limitation) or with $0.5 < S < 1$ (if there is a limitation).

4.2. Internals

The main task of storage vessel internals is to ensure an efficient charge and discharge operation.

4.2.1. Internals of Sliding-pressure Storage Vessels

Proper charging of sliding-pressure storage vessels by introducing steam into the vessel means to condense the steam with little or no temperature stratification and with low pressure drop. This is best done by creating a defined water recirculation in the vessel. In order to achieve water recirculation, steam could be blown into the vessel in its lower part. This, however, would mean a charge steam pressure considerably higher than storage pressure and, therefore, a pressure drop and exergy loss. On the other hand, blowing the steam into the upper (steam) part of the vessel would avoid any major pressure loss. But this would also mean heating up (by condensation) only the upper portion of the water and a consequent loss of storage capacity. The charging steam is therefore introduced in the upper part of the water volume (see Fig. 1.23). To create a defined water recirculation, guide plates are provided, thus separating the downcomer and riser zones. The riser zone is the one in which the charge nozzles are located. Their openings are directed upwards, thus producing a dynamic force on top of the buoyancy.

Proper discharging of the sliding-pressure storage vessel means to generate steam with low moisture from the water content of the vessel, keeping the steam cushion of the vessel as small as possible (for capacity reasons). This should, however, be sufficiently large to ensure almost complete steam/water separation by gravity. In sizing the vessel, the initial rise of the water level at the beginning of the discharge should be considered. Steam bubbles will develop in the upper part of the vessel and, subsequently, in the lower part also. The rise of the water level can, however, be greatly reduced by creating a defined water recirculation and its early start-up. The cylindrical guide plates provided for creating water recirculation during the charge operation are also used during discharge. They must be well below the minimum water level (which can be determined by using Fig. 2.3) in order to avoid hampering the recirculation. On top of these, frustroconical plates are arranged so as to concentrate the steam bubbles and in this way define the riser and downcomer zones during the beginning of discharge. Frustrocones pointing downward (Fig. 4.4) are preferable since in this case the heat from the wall, heat that is liberated during discharge, stimulates rather than impairs recirculation [4.5]. An intense recirculation minimizes the water level rise. This is important not only at the beginning of discharge but also later on in the process when the pressure has dropped and water content and specific water volume are low, for the specific steam volume increases and so does the tendency to splashing and foaming, thus requiring a larger steam cushion of the vessel for steam/water separation. Further moisture removal takes place in demisters arranged near the steam outlet in a manner similar to that in a boiler drum. In long horizontal vessels, a perforated horizontal discharge pipe may be used for demisting. A Venturi-shaped nozzle at the outlet will limit the discharge steam rate in extreme cases such as pipe rupture.

Besides equipment for charge and discharge, equipment for water level control is required for sliding-pressure storage. If the charging steam is saturated, then the amount of steam that will be discharged will be less than the amount charged, because of insulation losses. Accordingly, the water level will rise, and some draining will be necessary from time to time. If the charging steam is superheated, the water level will rise or fall, as the case may be. It will fall if the degree of superheating exceeds the heat loss per cycle. In this case either water is to be fed into the storage vessel or the charge steam temperature is to be reduced (for instance by use of a spray attemperator).

The charging of a sliding-pressure storage vessel with hot water discharge and external steam generation (expansion-type) may be performed either by filling with water and by applying steam (similar to sliding-pressure storage with internal steam generation), or by filling with hot saturated water (similar to displacement storage), or by simultaneously filling with water and steam (for

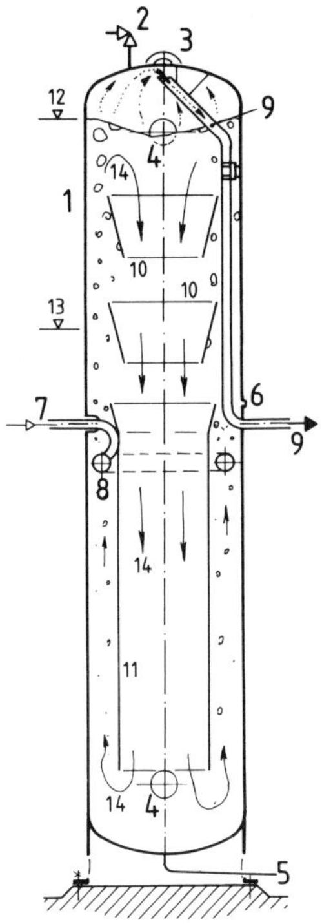

Fig. 4.4. Circulation during discharge of a sliding-pressure accumulator [4.5]
1 pressure vessel, *2* safety valves, *3* lifting lugs, *4* manholes, *5* blow-down line, *6* nozzle for level gauge, *7* charge line for steam, *8* ring-shaped charge nozzle, *9* discharge line for saturated steam, *10* frustroconical diverters, *11* cylindrical guide tube, *12* highest water level, *13* lowest water level, *14* sense of circulation during discharge

example, by preheating water in the upper part of the vessel). No internals are necessary for discharge.

4.2.2. Internals for Displacement Storage Vessels

The task of internals for displacement storage vessels is to avoid as completely as possible any temperature equalization between the hot and the cold parts. Temperature equalization means a loss of exergy and storage capacity; and, furthermore, the discharge temperature will decrease towards the end of discharge.

There are five mechanisms for temperature equalization:

(1) Mixing during charge and discharge operation.
(2) Axial heat conduction from hot storage medium in the upper part to cold medium in the lower part.
(3) Heat conduction in the pressure vessel wall and/or in the liner from the upper to the lower part.

(4) Thermal losses (surface losses) from the cylindrical part and from the ends to ambient.
(5) Mixing due to recirculation caused by surface losses. (In systems with low height–for instance, storage lakes–even instabilities may be caused by intense cooling of the uppermost layer.)

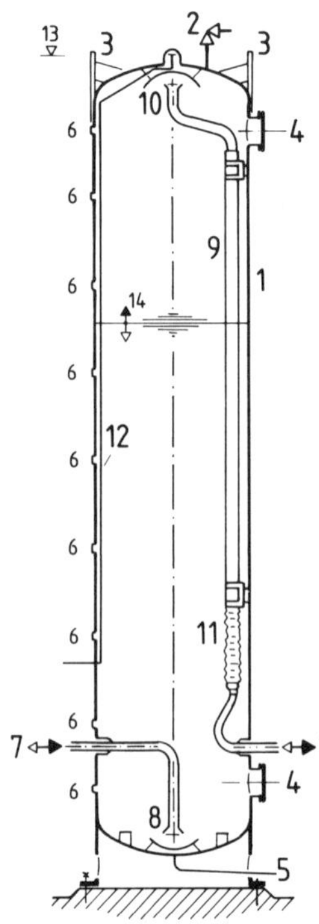

Fig. 4.5. Internals of a displacement storage vessel [6.21]
1 to *5* as in Fig. 4.4, *6* nozzles for temperature indicators, *7* cold water line, *8* lower flow distributor, *9* hot water line (insulated), *10* upper flow distributor, *11* compensator, *12* vent line, *13* water level, *14* transition zone

To reduce mixing losses according to (1) above, internals for reducing the inlet velocity and distributing the flow over the whole cross-section are provided. Fig. 4.5 shows internals of this kind for a vertical vessel [6.21], while Fig. 4.6 shows an unpressurized tank with a floating upper part which allows adaption to changes in water level [4.6].

Contrary to the mixing process, axial heat conduction in the storage medium according to (2) may be treated analytically [4.7]. The one-dimensional equation of heat conduction yields the local temperature distribution at any point in time. The temperature flattens with time. Fig. 4.7 shows the relative temperature

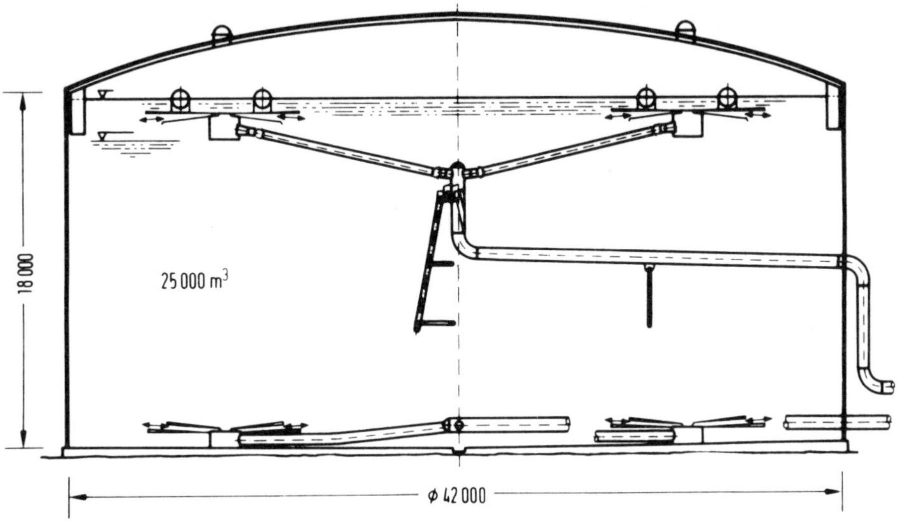

Fig. 4.6. Unpressurized displacement accumulator with floating upper charge/discharge equipment [4.6]–dimensions in mm

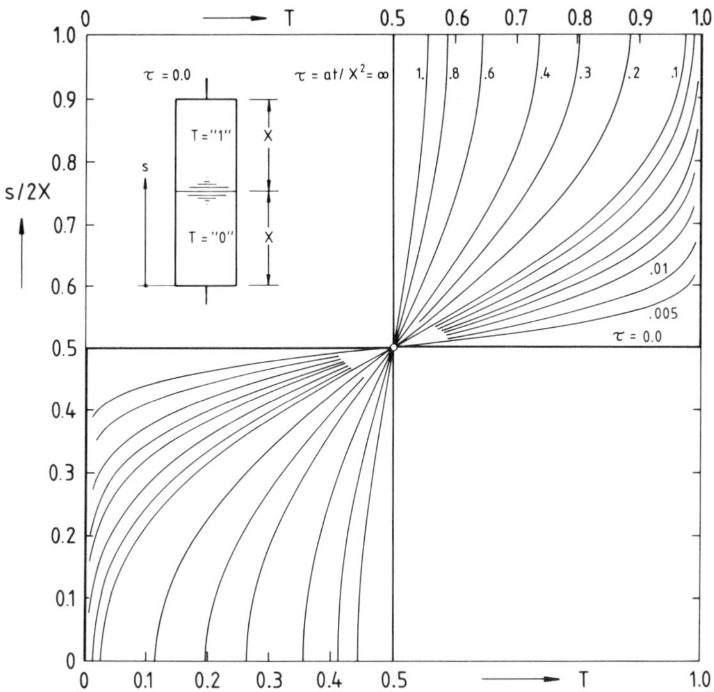

Fig. 4.7. Dimensionless temperature distribution versus dimensionless height of vessel (parameter: Fourier Number $\tau = at/X^2$) [4.7]

distribution over the dimensionless height of a vertical vessel with the dimensionless time (Fourier Number)

$$\tau = at/X^2 \tag{4.20}$$

as parameter where

$a = \lambda/\rho c$ temperature conductance (m^2/s)
t time (s)
X distance between vessel end and the center of transition zone (m).

For example, in a vertical vessel with $H = 12$ m, half-charged condition, $X = H/2 = 6$ m, operating temperature 260/60°C, mean water temperature 160°C ($\lambda = 0.69$ W/mK, $\rho = 910$ kg/m^3, $c = 4320$ J/kgK) $a = 0.69/(910)(4320) = (0.17)10^{-6}\,m^2/s$ and after a lifetime of the transition zone of 12 days ($t = 1\,036\,800$ s), τ becomes 0.005. If the transition zone is defined by the region where the temperature is between 250 and 70°C ($T = 0.95$ and 0.05), the height of the zone turns out to be $(0.575 - 0.425)H = 0.15H$ or 1.8 m.

Heat conductivity of steel (35 W/mK) is about 50 times that of water (0.69 W/mK). Therefore, despite the smaller cross-section, heat conduction in the pressure vessel wall is larger than that in the storage medium. For a vessel of 3 m diameter, a design pressure of 55 bar and 50 mm wall thickness, the cross-section ratio is about 0.06; the effective temperature conductivity is about four times that of water alone; and the time after which the temperature distribution according to $\tau = 0.005$ is reached actually turns out to be 3 days.

Thermally insulated separation plates, floating between hot and cold water, have been proposed [4.24]. These plates will suppress water conduction (and turbulence) but not wall conduction. As may be seen from Eq. (4.20), high vessels are another, more simple method.

4.3. Welded Pressure Vessels

Layout, design, inspection and operation of welded pressure vessels are all subject to national and international codes [4.4]. Annealing of welds is required at or above the following wall thicknesses: 30 mm according to the German Code and 38 mm (1.5 in.) according to the ASME Code.

Fig. 4.8 shows a vertical sliding-pressure storage vessel with hemispherical ends which was built for a steel mill in the United States and which permits equalization of steam from a waste heat boiler at a basic oxygen furnace (BOF) plant. The main data here are: $H = 19.2$ m; $D = 3.2$ m; $S = 6.0$; $V_i = 145$ m^3; $p_i = 60$ bar; the weight of vessel is: 150 tonnes* (this is almost a limit size for European railway profiles). The steam cushion in the charged condition is: 10%. Moreover, there are four support legs (one of them fixed), a platform, a ladder for inspection and a contact demister.

* 1 tonne = 1000 kg.

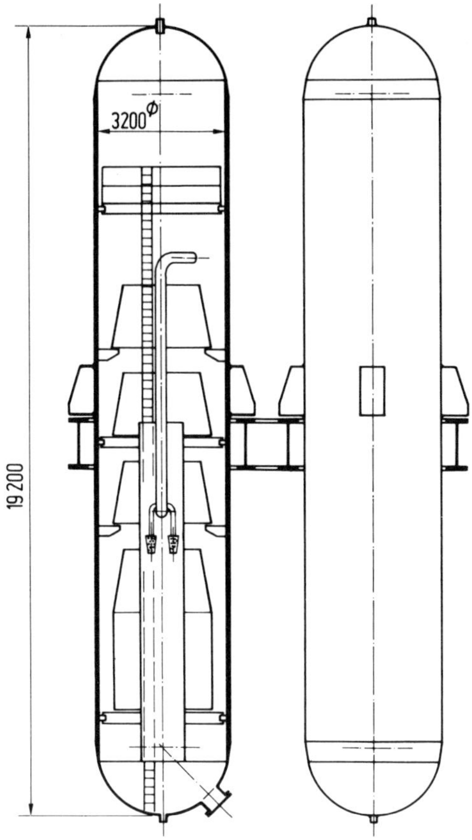

Fig. 4.8. Two vertical sliding-pressure accumulators, built for a US steel mill; $V_i = 145\ m^3$ each, $p_i = 60$ bar (Waagner-Biro AG)—dimensions in mm

4.4. Underground Pressure Vessels

Underground storage vessels [4.8] show some advantages in comparison to above-ground vessels:

- There seems to be practically no technical or economical size limitation.
- The cost may be lower than for above-ground vessels. There is a strong degression of specific cost with increasing size.
- The safety features are excellent.
- The area above the store may be used for plant purposes.

The disadvantages of underground storage vessels are:

- Special geological conditions must be met for low cost underground vessels.
- Inspection is more complicated.

– Hot underground vessels and vessels with temperature transients have not yet been built.

4.4.1. Pressurized Air Storage in Salt Domes

Salt domes are artificially produced by washing the salt out (Fig. 4.9) [4.9]. The pressure vessel produced in this way is, for all practical purposes, leak tight and may be used for pressurized air storage without requiring a liner. Salt entrainment during discharge should be accounted for in the design of discharge piping and turbine.

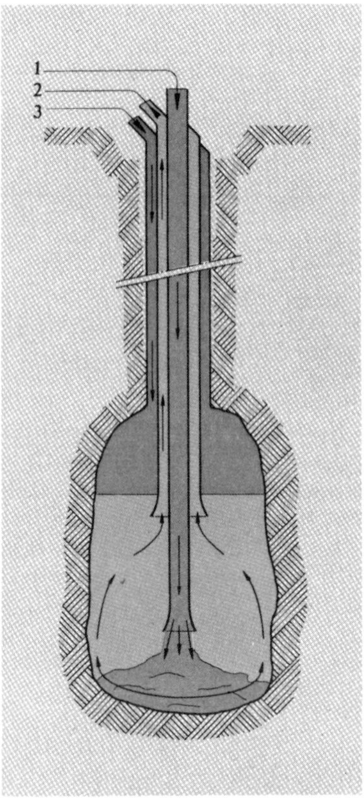

Fig. 4.9. Construction of salt domes by washing procedure [4.9]
1 washing water supply, *2* sole discharge, *3* inert fluid

The first store of this kind was built for a compressed air energy storage plant (see also Section 7.3.2 and Fig. 7.5) at Huntorf, in the Federal Republic of Germany. Table 4.1 gives the main data of the air storage vessels [7.6, 4.9].

Table 4.1. *Main Data of Pressurized Air Storage Vessels for the Compressed Air Energy Storage Plant at Huntorf* [7.6, 4.9]

Number of vessels	2
Type	salt dome (cavern)
Total volume	(2)(150 000) = 300 000 m^3
Storage capacity	(290 MW)(2 h) = 580 MWh
Depth (top/bottom)	650/800 m
Diameter	max. 60 m
Salt thickness above caverns	min. 100 m
Salt thickness between caverns	min. 160 m
Ovality of cavern	max. 2:1
Type of storage	sliding-pressure (air)
Working pressure (upper/lower)	65/45 bar
Maximum allowable pressure	100 bar
Maximum speed of depressurization	10 bar/h
Maximum speed of pressurization	2.5 bar/h
Maximum air temperature	50°C

4.4.2. Hot Water Displacement Storage in Rock Caverns

Fig. 4.10 shows the hot water displacement storage in rock caverns as proposed by Margen [4.10] for peak power generation from nuclear power plants. Hot pressurized water of 217/73°C and about 30 bar is stored 60 m below ground in a thin steel tank that is thermally insulated on the outside. A compressor is used to keep the air between the steel tank and the rock cavern (which has been cleaned with concrete) at the pressure corresponding to water pressure in the upper part of the tank. Pressure equalization is ensured by a flow of water from the tank to a pressure equalizing vessel from which the water is piped to the degasser of the hot well. In this way, the tank is designed only for its own hydrostatic pressure. In order to avoid steaming of the hot water discharge flowing upward, the hot water in the storage tank is pressurized.

This technology was further investigated in a Canadian study (see Fig. 4.11, Table 4.2 [4.11]) and was found to be feasible.

Table 4.2. *Main Data of Hot Water Displacement Storage in a Rock Cavern, Canadian Study* [4.11]

Number of vessels	6
Total water volume	(6)(15 000) = 90 000 m^3
Total cavern volume	205 000 m^3
Storage capacity	(417 MW)(8 h) = 33 000 MWh
Type of storage	displacement (water)
Water temperature (upper/lower)	265/95°C
Saturation temperature	51 bar
Air temperature, interspace	65°C
Storage depth	550 m
Water/air pressure	95 bar

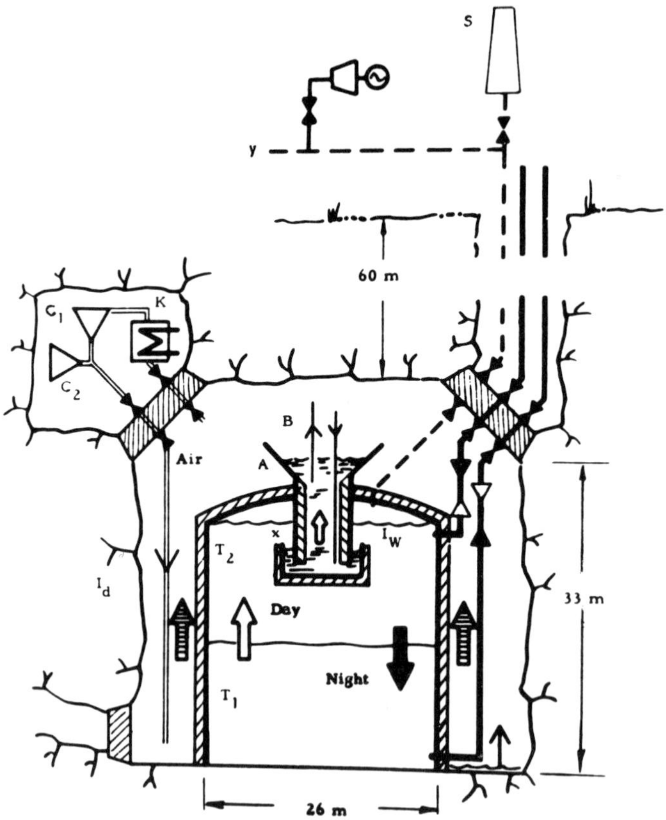

Fig. 4.10. Hot water displacement accumulator in rock caverns, proposed by Margen [4.10]
A narrow sealing passage, *B* pressure balance tank, C_1 air circulator, C_2 air compressor, I_d dry insulation, I_w wet insulation, *K* air cooler, *S* stack, *x* sealing flow to prevent intrusion of gas, *y* steam pressurizing supply

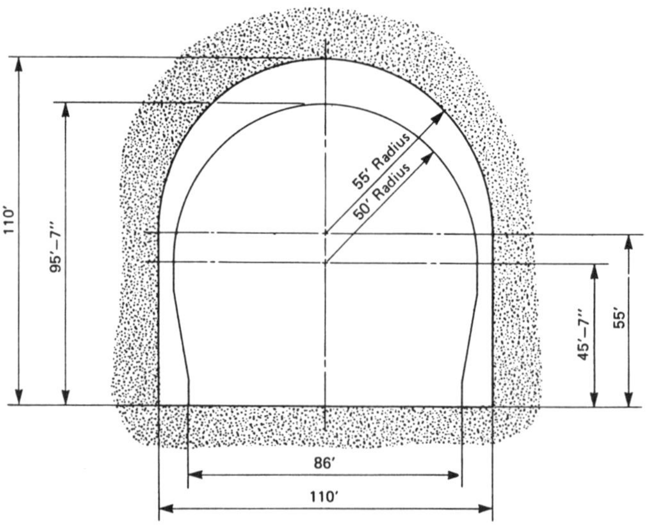

Fig. 4.11. Vessel shape and cavern cross section of a hot water displacement accumulator in a rock cavern, Canadian study [4.11]

4.4.3. Hot Water Sliding-pressure Storage in Rock Caverns

Transmitting the storage pressure via pressurized air to the cavern is not feasible for sliding-pressure storage. In a United States study [4.8], a method of transmitting the forces via 12 in. of high temperature, high strength concrete from liner to cavern is described. Fig. 4.12 shows the wall of the spherical pressure vessel. The steel liner is tied to the rock by bolts of about 1 m length; it is welded on site. There is heat flowing into the rock, developing a temperature profile that serves as thermal insulation. The higher insulation loss during the first few years of operation may be considered as a heat investment. For a sphere of about 100 m diameter, the heat investment is comparable to one heat content of the rock chamber [4.8].

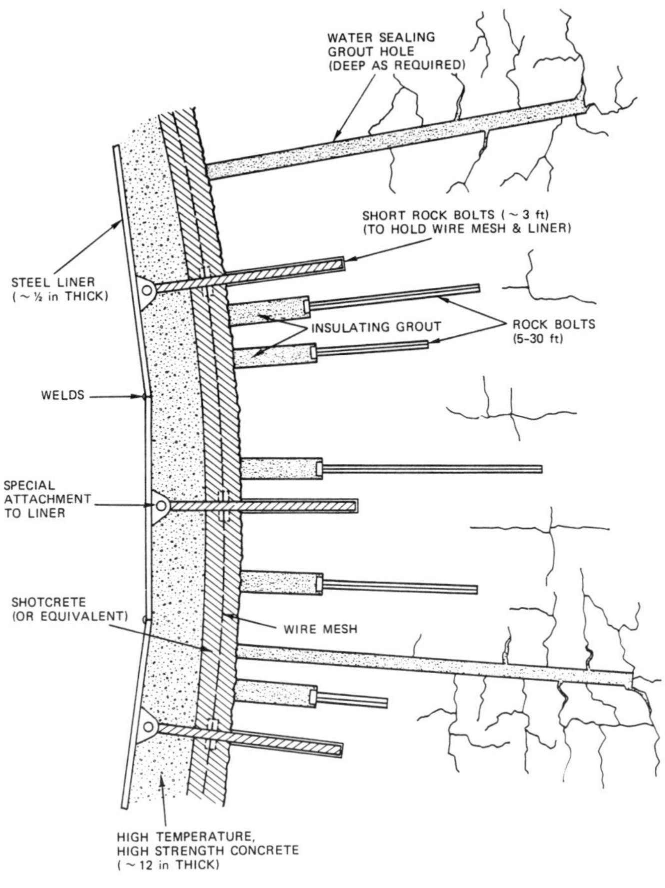

Fig. 4.12. Cavity wall of an underground sliding-pressure accumulator [4.8]

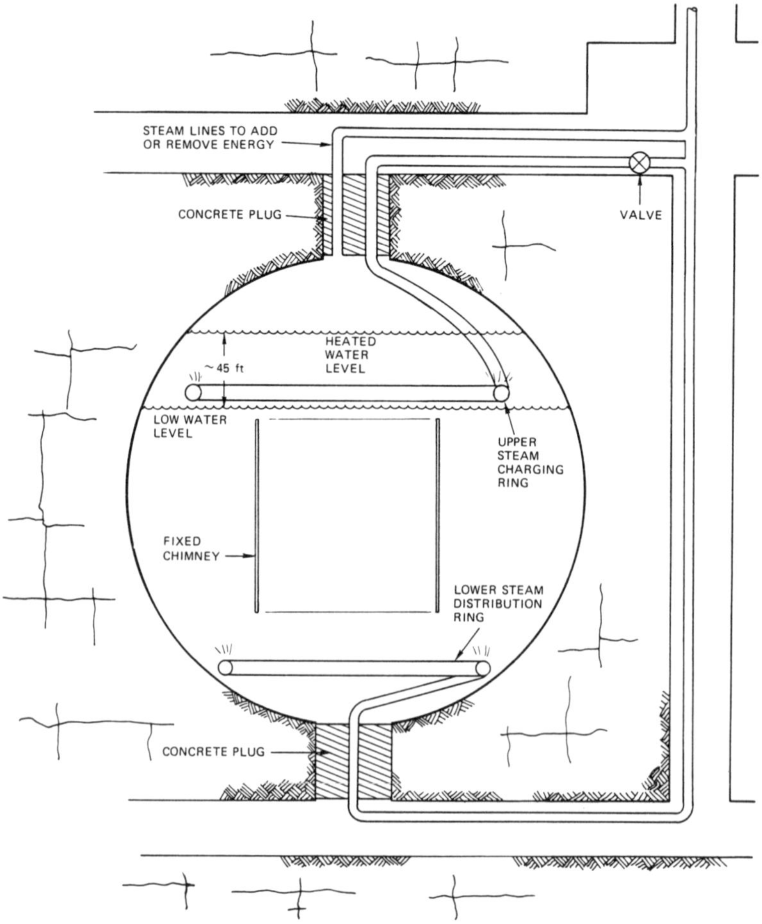

Fig. 4.13. Equipment and internals of the underground sliding-pressure accumulator [4.8]

Fig. 4.13 shows the completed underground sliding-pressure store. Two steam distribution rings are provided in order to minimize pressure loss during charging; the lower one is acting at the beginning of the charging operation and the upper one towards the end.

A charge pressure of 75 bar and a discharge pressure of 40 bar are suggested in connection with nuclear power plants of the PWR and BWR type and 130/75 bar for fossil fired steam power plants, that is, the temperature cycling amounts to 40 K in both cases. Exergy density is low (25 or 29 kWh/m^3, respectively) and the required storage volume is therefore high, although storage efficiency is high (about 90%).

Fig. 4.14 shows the general arrangement of such a plant. The cavern is located at such a depth that lithostatic pressure is 1.33 times the maximum storage pres-

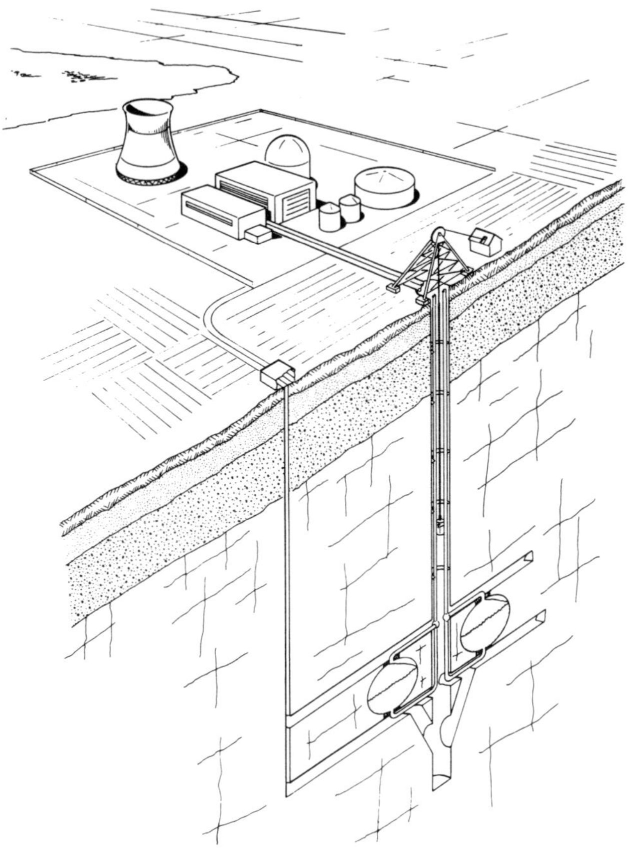

Fig. 4.14. General arrangement of a nuclear power plant with underground sliding-pressure accumulators [4.8]

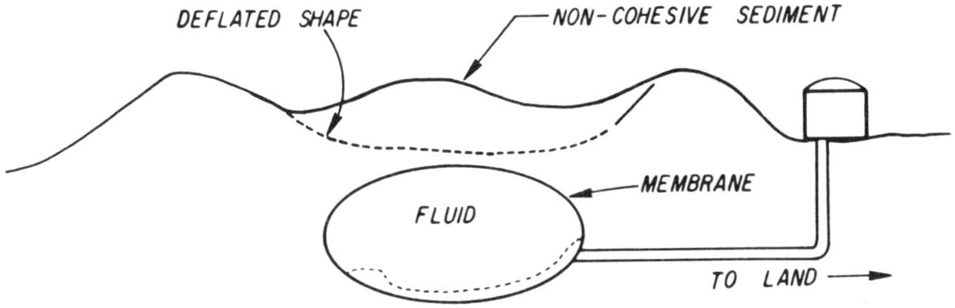

Fig. 4.15. Buried flexible accumulators [4.12]

sure. This ensures compression at the cavern boundaries and avoids crack propagation. For 75 bar = $(7.5)10^6$ Pa maximum pressure and a rock density of 2500 kg/m^3, the depth becomes $(1.33)(7.5)(10^6)/(2500)(9.81) \doteq 400$ meters.

4.4.4. Flexible Containers

It has also been proposed [4.12] that hot water stores be located below the sea or in non-cohesive beds (dry sand in deserts), by using hydrostatic pressure for pressure equalization. The pressure vessel in this case consists of a flexible membrane with a suitable thermal insulation (which is also to be flexible) at its inside or outside (see Fig. 4.15).

Proper support will be required in the sea. Because of the lower density of sea water and dry sand as compared to rock, the container will have to be buried even deeper than the rock caverns described above.

4.5. Prestressed Pressure Vessels

4.5.1. General

Prestressed pressure vessels differ from welded pressure vessels in that different specialized elements are provided for the different functions:

- The dimensional stability is ensured by the vessel wall which is essentially under compression and is either homogeneous or consists of ring-shaped or sector-shaped parts.
- The prestress load is applied by tendons consisting of steel cables which are essentially under pure tension.
- The internal pressure reduces the compression stress in the vessel wall and somewhat increases tension stress in the tendons.
- The sealing function is performed by a liner, since the vessel wall itself may not be leak-tight or it may not be resistant to the storage medium (indeed, it may be neither).

The required properties of these elements are therefore:

- Vessel wall: resistance to compression at working temperature, dimensional stability.
- Tendons: high yield strength at working temperature; low relaxation.
- Liner: high elasticity (low Young's modulus, high ductility); resistance to the storage medium.

Thermal insulation is a further element. Here, there are several concepts which, to a great extent, determine the design. They are as follows:

(a) *Cold Liner*

The thermal insulation is inside the liner and is in contact with the storage medium. Inspection of the liner is difficult.

(b) *Hot Liner*

The thermal insulation is between the liner and the vessel wall and must therefore withstand the internal pressure. The liner has to take up differential thermal expansion and therefore has to be extremely ductile.

(c) *Hot Liner/Hot Vessel Wall*

The thermal insulation is between vessel wall and tendons and must therefore be pressure resistant. The vessel wall should have high thermal conductivity and high fatigue resistance. Moreover, the vessel wall is prestressed by thermal expansion during heating up.

(d) *Hot Liner/Hot Vessel Wall/Hot Tendons*

The thermal insulation is beyond the tendons, similar to the situation in a welded pressure vessel. High strength and low relaxation of the cables at working temperature is required.

Prestressed pressure vessels may also be classified according to the material of the vessel wall. According to such classification, we have

- Prestressed Concrete Pressure Vessels (PCPV);
- Prestressed Cast Iron Vessels (PCIV).

Since concrete is not resistant to temperatures much above ambient, only concepts (a) and (b) are applicable to the PCPV. On the other hand, concepts (a) through (d) are, in principle, applicable to the PCIV.

The main advantage of the prestressed designs as opposed to the underground pressure vessels is that they are situated above ground and therefore do not require special geological conditions except for a certain strength of the subsoil which must take foundation pressure. Like welded pressure vessel (and unlike underground vessels), they are suitable for almost any site.

The advantages of the prestressed design as contrasted to welded pressure vessels are:

- Possibility of easy construction on site and hence suitability for very large volumes.
- Improved safety because of the redundancy of the many independent tendons, as well as because of the tendency to gapping with pressure reduction at overpressure.
- Higher technological limits of pressure and volume.

4.5.2. Prestressed Concrete Pressure Vessel (PCPV)

Based on experience with PCPV for gas-cooled reactors [4.13], there are codes for the design of PCPV available in the United States [4.14], in Great Britain [4.15] and in France [4.16]. Stress analysis of PCPV is rather involved: The theory of thin elastic shells does not normally apply, transition zones—for instance, from cylindrical part to flat end—are difficult to assess. Model testing is therefore mandatory. The ASME Code [4.14], for example, requires 1:4 scale models for long-term behaviour and 1:14 scale models for elastic and fracture behaviour.

Stresses are classified according to load categories. According to the ASME Code, the following limits apply to the "Normal Category Load Combination":

- Concrete Body: Primary compressive stress is to be less than $0.3C$ times compression strength, where the value of C ($\leqslant 1$) depends on whether the stress is one-dimensional, two-dimensional or three-dimensional. Primary tension is not permitted. The allowable maximum temperature of the concrete body is 65°C (150°F) and of "hot spots" 120°C (250°F).
- Cables: Axial tensile stress is to be lower than 0.7 tensile strength and 0.8 yield strength. The allowable maximum temperature is 65°C.

Basic design philosophy is that the concrete shall withstand 1.2 times the working pressure and, at the same time, 1.5 times the calculated temperature gradient in the wall. Fracture of the cables shall occur at not less than 2 to 2.5 times the design stress.

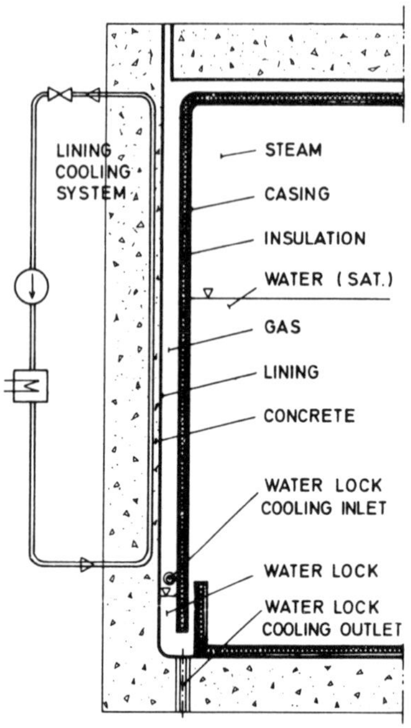

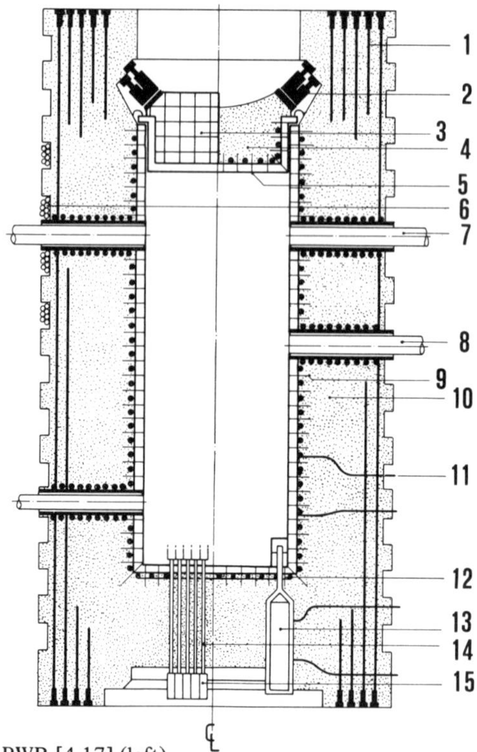

Fig. 4.16. Swedish concept of a PCPV for a BWR [4.17] (left)

Fig. 4.17. German concept of a PCPV for a BWR [4.18] (right)
1 axial tendons, *2* support elements, *3* steel cover, *4* concrete cover, *5* liner, thermal insulation, *6* tangential tendons, *7* steam outlet, *8* feed water inlet, *9* bolts, *10* concrete, *11* liner cooling system, *12* bottom cooling system, *13* recirculation pump, *14* control rod guide tubes, *15* control rod drive

Considering the low allowable concrete temperature, the relatively low thermal conductivity and the high wall thickness, only the concepts of cold vessel wall are possible. Moreover, even though there is no heat generation in

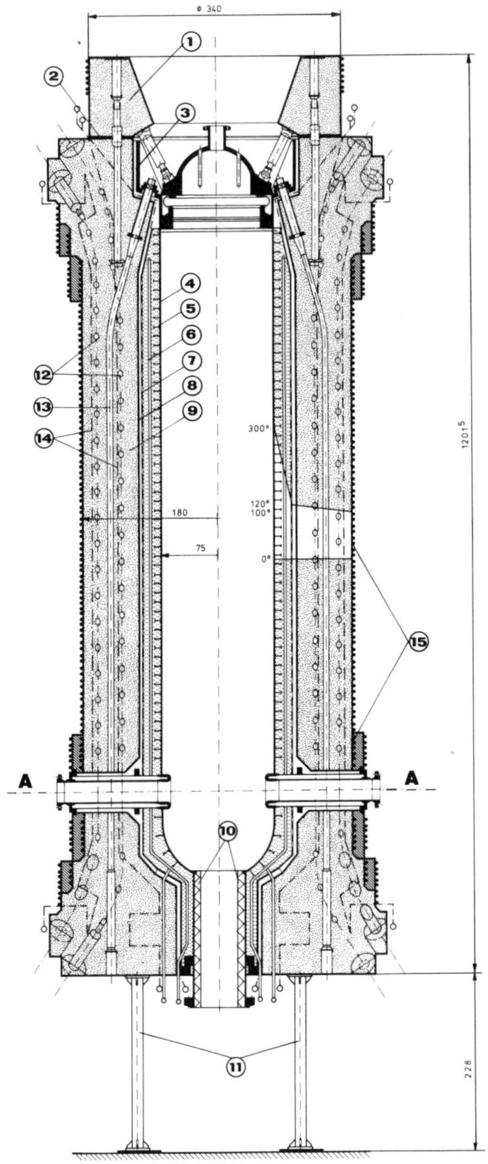

Fig. 4.18. Austrian (Seibersdorf) concept of a PCPV with hot liner [4.19], test vessel–dimensions in cm

1 prestressed concrete ring, *2* interspace, *3* support ring, *4* hot liner, *5* thermally insulating concrete, *6* leak tightness control, *7* inner cooling system, *8* interspace, *9* concrete, *10* internal insulation, *11* support columns, *12* helical tendons, *13* axial tendons, *14* outer heating/cooling system, *15* tangential tendons

the wall as in nuclear pressure vessels, cooling systems for the concrete are necessary.

The conditions of a PCPV hot water storage vessel are similar to the conditions of BWR or PWR vessels as regards the pressure, the temperature and the kind of medium. PCPV for BWR and PWR have not yet been built, although there are a number of concepts, the main difficulty being the design of the liner and the thermal insulation. Some of them are mentioned below:

- *The Swedish Concept for BWR* (Fig. 4.16) [4.17]: This concept features a cold liner and a cylindrical interspace with pressurized nitrogen, similar to the air interspace of Fig. 4.13. The thermal insulation of the lower flat end is performed by several layers of stagnant water.
- *The Krupp Concept for BWR* (Fig. 4.17) [4.18]: Here, there is a hot liner as well as a pressurized thermal insulation consisting of ceramic spheres in a binder.
- *The Seibersdorf Concept* (Fig. 4.18) [4.19]: In this concept, there is a hot liner with low thermal expansion coefficient as well as a pressurized thermal insulation consisting of porous concrete (LBn 450) with low thermal conductivity and a main body of temperature resistant concrete which is able to operate at somewhat elevated temperatures of 100 to 120°C or more.

The application of such reactor-vessel concepts to hot water storage means a simplification (less penetrations) on the one hand. On the other hand, however, the number of temperature and pressure changes will be much higher than in a reactor and questions of thermal insulation with cold liner and of fatigue of a hot liner will be even more important.

4.5.3. Prestressed Cast Iron Vessel (PCIV)

In the late 1960s the Prestressed Cast Iron Vessel (PCIV) concept was proposed and developed by Siempelkamp Giesserei KG of Krefeld, in the Federal Republic of Germany. This concept utilizes solid or hollow cast iron blocks of suitable size (machined as necessary) with keys to ensure their relative position. In addition to the advantages of prestressed designs in general, a number of advantages are claimed [4.20–4.25]:

- Cast iron has a high compressive strength (about 1000 N/mm^2; that is, it is 20 times that of concrete), leading to low wall thickness and weight and low foundation cost.
- Cast iron is suitable for high temperatures (that is, it does not require a wall or liner cooling system).
- The PCIV is erected from prefabricated blocks of optimum size and weight and may be easily disassembled and re-erected if necessary.

The PCIV was first developed as a reactor-vessel alternative for the German High Temperature Reactor THTR of 300 MW. In 1972, a 1:7.5 scale model was built and tested (Fig. 4.19). The first commercial PCIV was a high-pressure energy storage vessel which acts as control-rod-drive pressure vessel for the THTR (Figs. 4.20 and 4.21). Its main data are as follows [4.25]:

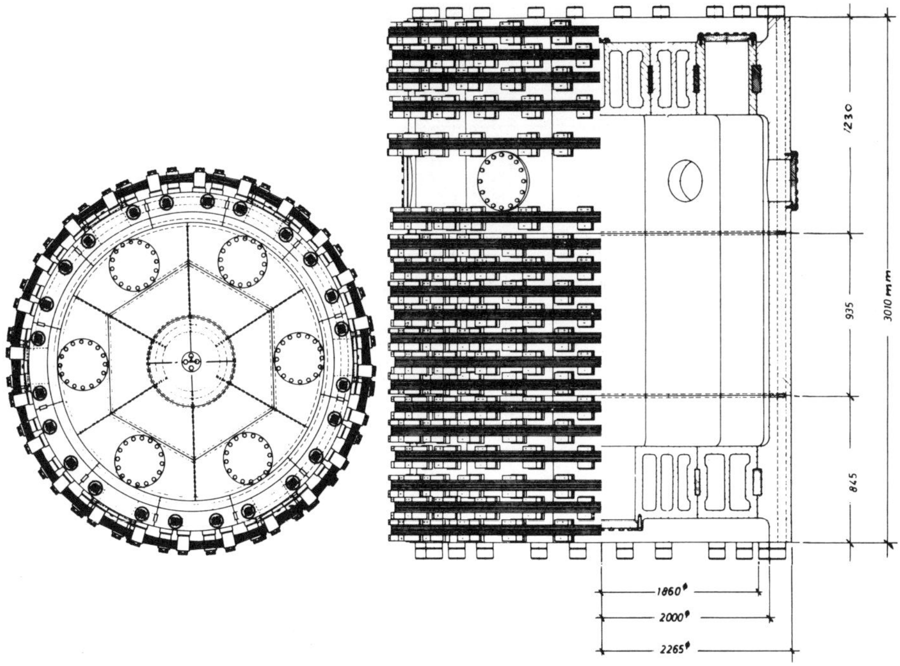

Fig. 4.19. PCIV test vessel for 90 bar, built by Siempelkamp in 1972 [4.20]

Fig. 4.20. PCIV according to Fig. 4.21 leaving the factory (courtesy of Siempelkamp + Steinmüller GmbH, Krefeld)

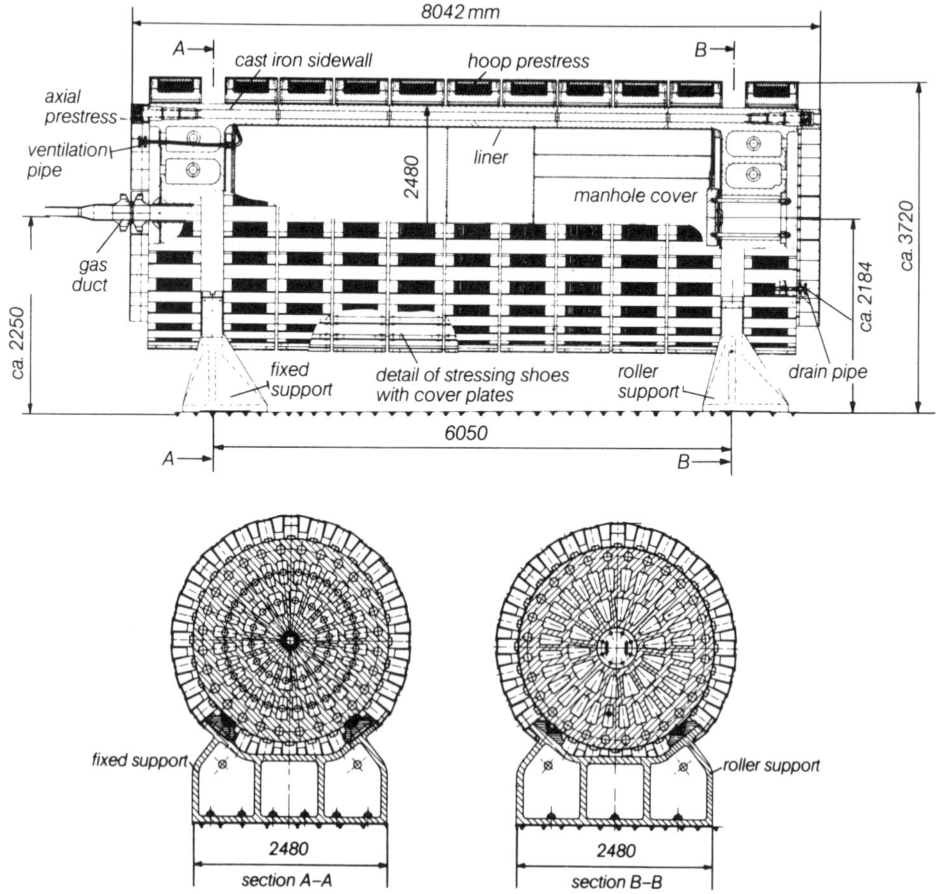

Fig. 4.21. PCIV for 230 bar helium control rod drive (compressed gas energy store) of the THTR plant at Schmehausen, FRG [4.25, 4.32]

Storage Medium:	Helium
Operating Pressure:	230 bar
Internal Volume:	17.5 m^3
Overall Weight:	200 tonnes
Cast Iron Type:	with lamellar graphite
Cables:	St 1670 (1470 N/mm^2 yield strength, 1670 N/mm^2 ultimate strength), 7 mm diameter.

Larger PCIVs have been designed with lamellar and spherical graphite cast iron and have been proposed for air and pressurized hot water storage. A PCIV pressurized air storage vessel was proposed for the quick start-up of a 100 MW gas turbine (see Section 7.3.3 and Fig. 7.7).

Fig. 4.22 shows a vertical PCIV of 2000 m^3 and 60 bar with cold liner, designed as a sliding-pressure storage vessel, while Fig. 4.23 exhibits a horizontal PCIV which consists of prefabricated rings that are tangentially prestressed in

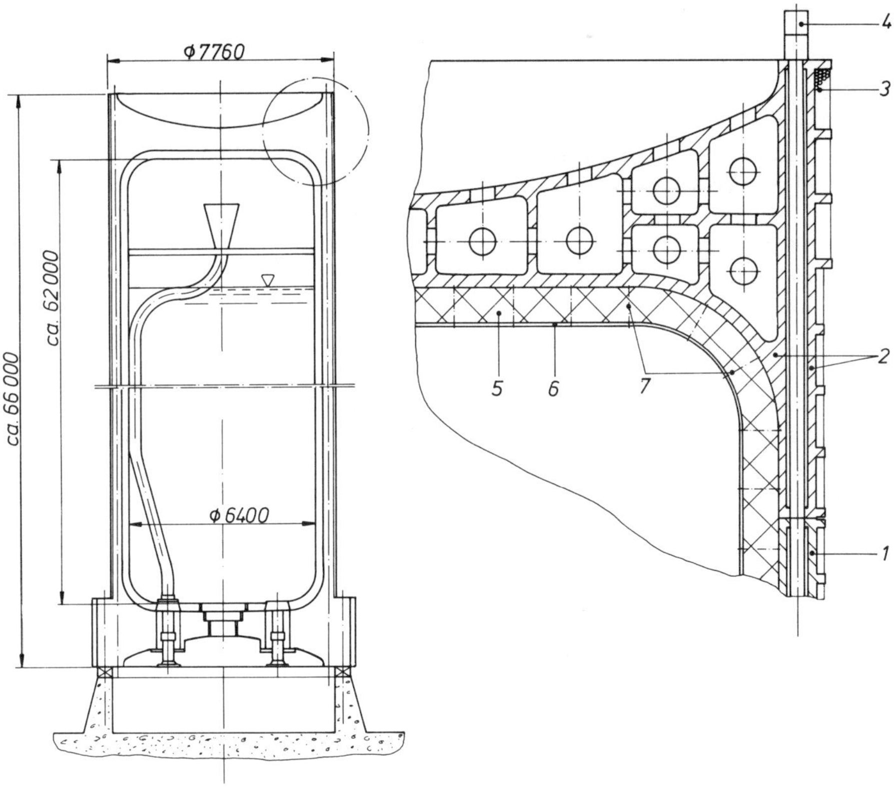

Fig. 4.22. Vertical PCIV for sliding-pressure storage, 2000 m^3, 60 bar (courtesy of Siempelkamp Giesserei GmbH & Co., Krefeld)–dimensions in mm
A general arrangement, *B* detail of head: *1* cast iron block of cylindrical wall, *2* cast iron block of head, *3* tangential tendons, *4* axial tendons, *5* thermal insulation, *6* liner, *7* liner support

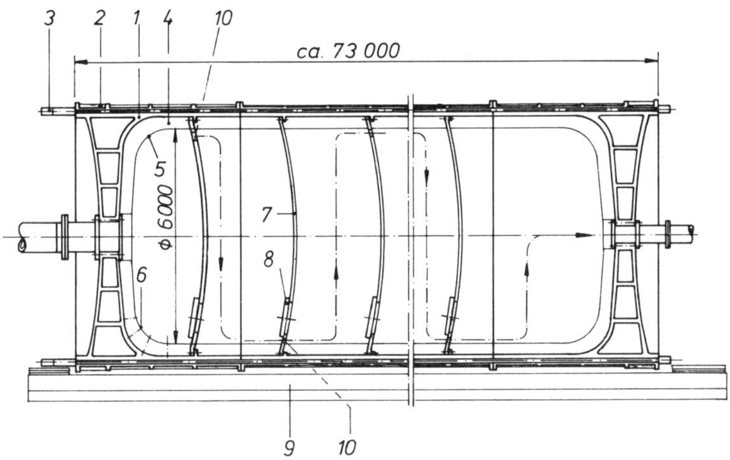

Fig. 4.23. Horizontal PCIV for displacement hot water storage, 2000 m^3, 60 bar (courtesy of Siempelkamp Giesserei GmbH & Co., Krefeld)–dimensions in mm
1 cast iron blocks, *2* tangential tendons, *3* axial tendons, *4* thermal insulation, *5* liner, *6* liner support, *7* baffles, *8* manhole cover plates, *9* support plate, *10* openings for water flow

the shop and are connected by axial tendons to a modular storage vessel whose length (and capacity) may be selected on basis of the individual needs. It is designed as displacement-type storage vessel. Mixing is suppressed by means of a labyrinth system similar to that shown in Fig. 4.25. PCIVs with unit volumes of up to 8000 m^3 have been designed and proposed [4.23].

The penetrations are located in the heads. Apart from the main penetrations for charge and discharge pipe or pipes, there may be several smaller penetrations for a safety valve, pressure transducers, drain pipes, etc. Preferably, these should be combined with a charge or discharge pipe penetration. Coaxial pipe penetrations may be utilized in order further to reduce the number of penetrations. The penetrating pipes are connected to the cast iron body by means of an internally situated flange which also takes care of possible liner movements. Large pipes may be provided with a Venturi-shaped insert in order to limit the flow rate in case of pipe failure.

4.6. Unpressurized Vessels

Liquid storage media that do not require pressurization at working temperature (for instance, water below 100°C or "high boiling" liquids) are stored in unpressurized vessels. Their internal pressure is mainly given by hydrostatic pressure and possibly by a slight overpressure of an inert gas cushion above the liquid.

Optimum shape has already been discussed in Section 4.1.4. The cost of the thermal insulation, the liner (if any) and the corrosion protection, is relatively more important than for pressure vessels. Therefore, classification is made according to storage medium and temperature.

4.6.1. Unpressurized High-temperature Vessels

Above-ground tanks (made of stainless steel if necessary for reasons of corrosion) are common for unpressurized high-temperature vessels. Molten salt, oil and liquid metals do require an inert gas atmosphere. Moreover, the vessel is usually slightly pressurized so as to avoid the ingress of air.

As an example, Fig. 4.24 shows a latent heat store that was proposed by Boeing for solar gas turbine plants [2.30]. This proposal was advanced as an alternative to pressurized solid store. The storage medium is a eutectic salt (63 NaF/37 ZnF_2) which has a melting temperature of 640°C. It is contained in a quadratic concrete vessel of 722 m^3, the vessel being thermally insulated inside and sealed by a liner of Hastelloy-N. Storage is indirect. Heat is transferred to and from the working medium (helium of 34 bar) via tubes. The store permits operation of the solar gas turbine of 50 MW electric power for 6 hours without insolation.

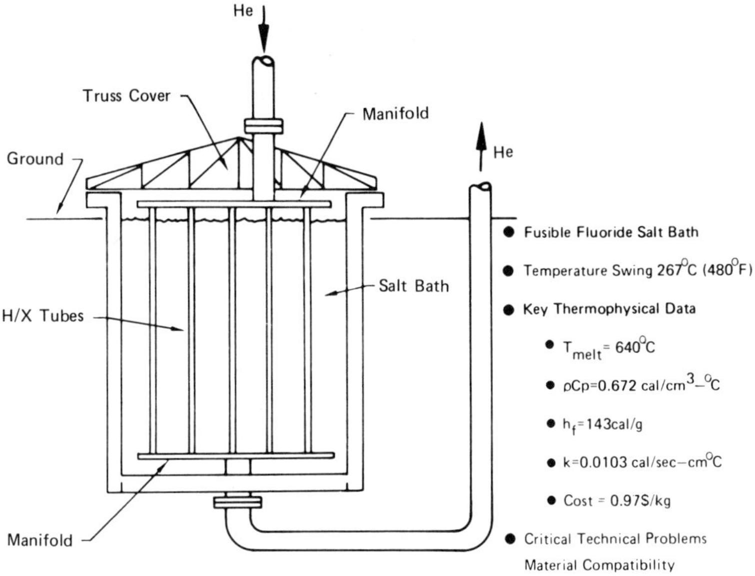

Fig. 4.24. Latent heat store proposed by Boeing [2.30]

4.6.2. Unpressurized Low-temperature Vessels

Water is the most common storage medium for low-temperature storage. For chilled water storage (as required in air-conditioning plants) the temperature range is about 5 to 15°C; for storage of hot water (for heating and domestic use) the range is about 30 to 95°C.

The temperature and density difference of chilled supply and return water is rather small. However, if mixing is to be avoided, then the concepts of two-vessel storage (see Section 2.3.2a), of multi-vessel storage (Section 2.3.2b), or of single-vessel storage with floating separating disc may be considered. If some

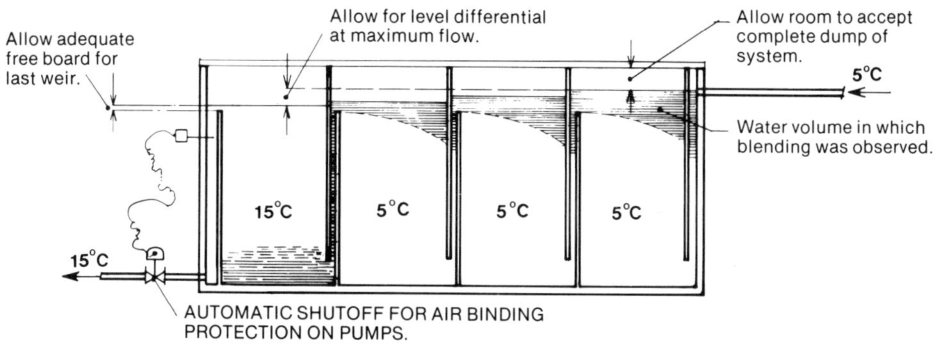

Fig. 4.25. Displacement accumulator with labyrinth concept [4.26]

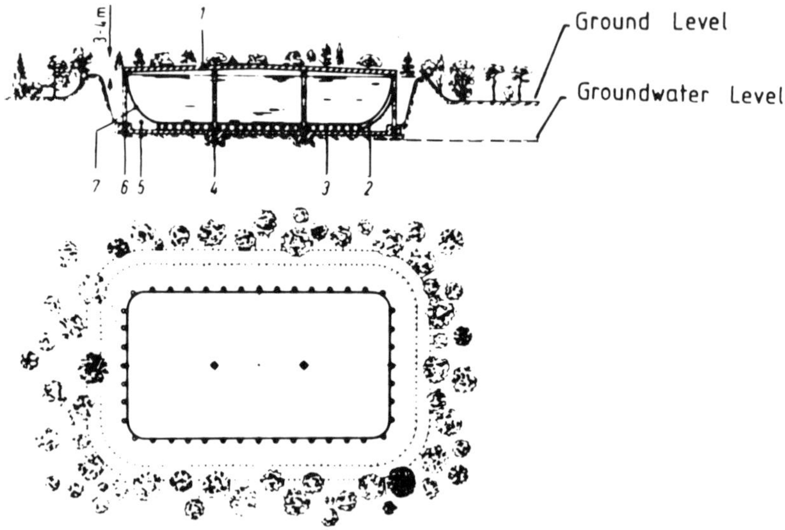

Fig. 4.26. Steel membrane reservoir [4.27]
1 insulated roof, *2* side wall stiffening, *3* bottom support, *4* roof pillars, *5* thermal insulation, *6* sidewall pillars, *7* side wall

mixing is allowed, the labyrinth concept (Fig. 4.25) is applied [4.26]. Mixing must also be considered for hot water storage vessels (which usually show a low H/D ratio and large mixing areas) as well as for the steel membrane reservoir shown in Fig. 4.26 [4.27, 4.30].

Fig. 4.6 shows a hot water storage tank of 25 000 m^3 and $H/D = 0.43$ that was converted from a former oil tank [4.6]. It is being used in a district heating system. The supply pipe floats to follow water level variations. In the upper part, steam with low overpressure (0.003 bar) avoids the ingress of air. Cost of newly

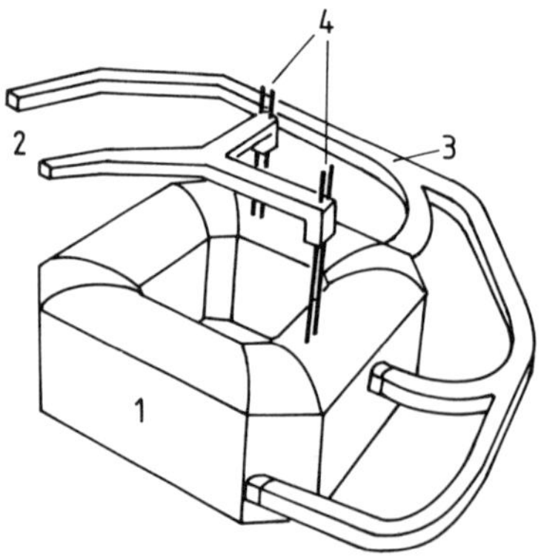

Fig. 4.27. Rock cavern for hot water storage at Lyckebo, Sweden (near Uppsala), 100 000 m^3, 90/40°C [4.28]
1 rock cavern, *2* access port, *3* tunnel used for excavation, *4* charge and discharge piping

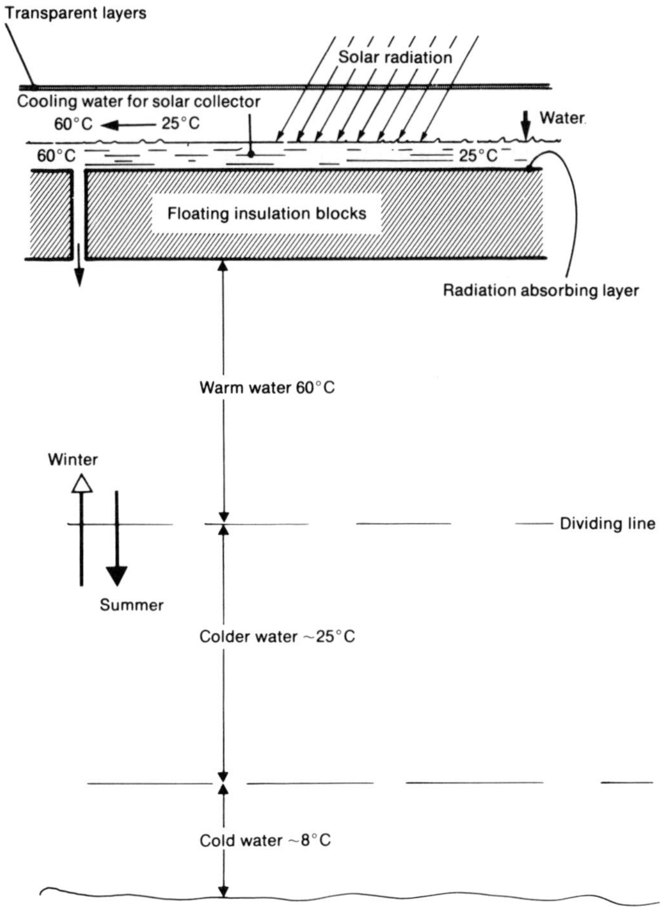

Solar collector combined with heat storing.

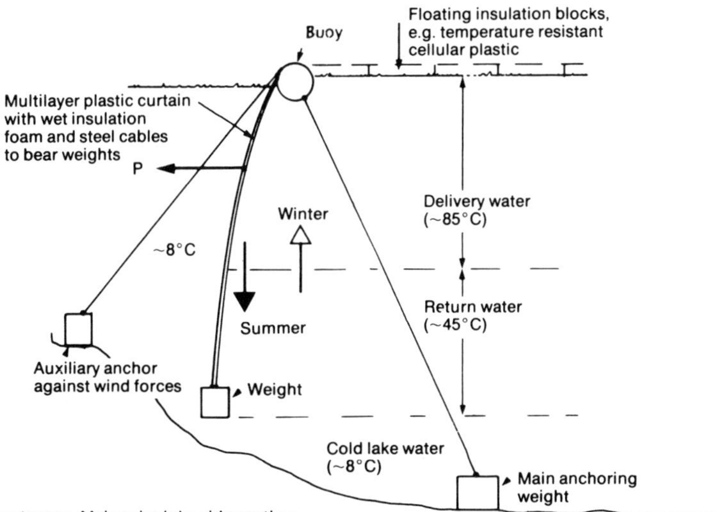

Lake storage. Main principle, side section.

Fig. 4.29. Swedish concepts of lake storage [4.31]

built vessel of this kind is quoted as 107 DM/m^3 (1979)* for medium and 58 DM/m^3 for very large vessels [6.20].

Fig. 4.27 shows a rock cavern which was constructed at Lyckebo (north of Uppsala, Sweden). Built for seasonal solar energy storage and containing 100 000 m^3 of water, the cavern is ring-shaped and is not thermally insulated [4.28]. Working temperature is 90/40°C. To a certain extent, the innermost layers of rock take part in storing heat.

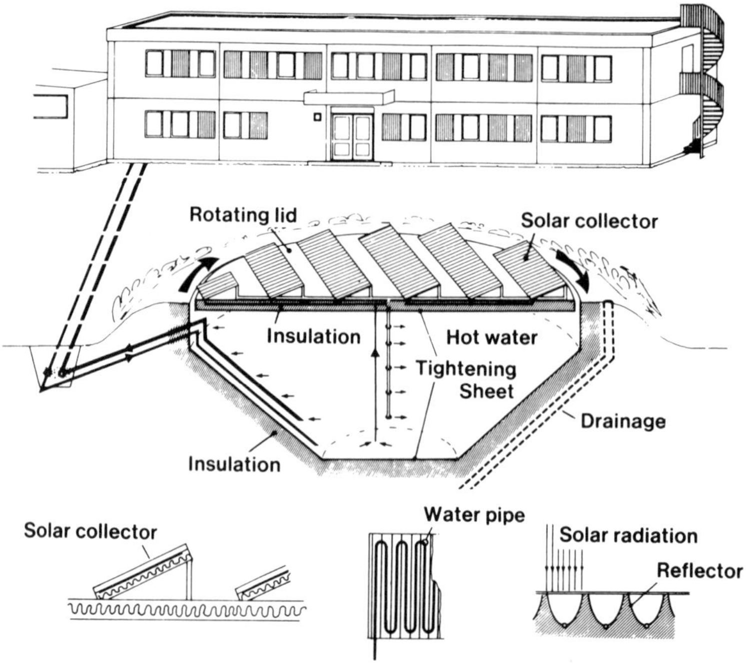

Fig. 4.28. The prototype solar earth pit storage at Studsvik, Sweden, 640 m^3, 70/30°C [4.27, 4.29]

Rock caverns (and open pits) which are partly filled with rock have also been suggested. Construction is somewhat cheaper and stratification of water is improved in this way.

Further proposals such as earth pit reservoirs (see Fig. 4.28) [4.29], artificial lakes with floating cover [4.30] and membranes floating on a lake (see Fig. 4.29) [4.31] are not vessels in the strict sense anymore. Further methods of low temperature thermal energy storage may even lack well-defined borders such as aquifers with hot water storage and ground water, subsoil and gravel storage for heat pumps, etc.; these methods are described in Section 6.7.

---

* 1 DM (1979) = 0.54 \$ (1979).

5. Thermal Energy Storage in Industry

5.1. General

Thermal energy storage (TES) has been used in industry for many decades [5.1, 5.2]. Furthermore, improved energy conservation has brought about new applications. Although TES alone does not "save" energy in the strict sense by itself–on the contrary, it will lead to exergy and sometimes energy losses (see Chapter 3)–it will greatly alleviate industrial energy management and will, accordingly, reduce external energy demand in many instances.

In a way that is similar to the situation with power plants, there are three tasks of TES:

(a) peak heat load coverage; for example, for discontinuous heat demand or for steam generator reserve duty;
(b) equalization of discontinuous heat sources; for example, the utilization of waste heat from discontinuous processes (batch operation);
(c) peak power and heat load coverage in industrial combined heat and power (CHP) plants.

Moreover, there is the quasi-continuous operation of regenerators for combustion air heating.

More recent applications include waste heat utilization from flue gases of basic oxygen furnaces (BOF) or electric arc furnaces and dry cooling of incandescent coke. In these instances, there is also an ecological effect to be considered, that is, reduction of air pollution and reduction of water consumption for cooling, quenching, etc.

Table 5.1 shows that mainly solid high temperature storage and steam/hot water storage for temperatures of up to 300°C are used for industrial applications [2.15]. Some typical applications are described in the following discussion.

In the pulp and paper industry, TES will be important for purposes of achieving constant full load of boilers which are fired with wood chips, barks, etc., and whose load following capability is rather poor. Another typical application is steam storage in steam networks in which paper mills are the main users of steam. In case of paper rupture (which may occur several times a day) steam would have to be blown off for 5 to 10 minutes. However, with TES the excess steam is recovered.

Table 5.1. *Industrial Processes Suitable for TES*

	Process Diversion	Energy Savings	Temperature °C	TES Function	TES Technology
Iron and Steel					
Combustion air preheat	Sm	L	500–1350	S	Regenerator
Steam generation from combustion products	L	L	250	S	Steam acc.
Waste heat electric arc furnaces	Sm	L	500–1650	S	Regenerator
Waste heat from BOF	Sm	L	up to 2000	S	Steam acc.
Dry coke cooling	L	L	1000	S	Packed bed
Paper and Pulp					
Waste fuelled hog boilers	L	L	250	S	Steam acc.
Paper breaks	Sm	L	120–150	S	Steam acc.
Food Processing					
Washing	L	U	60	D	Hot water
Scalding	L	U	50–60	D	Hot water
Pasteurization	L	U	65	D	Hot water
Blanching	L	U	85	D	Hot water
Cooking	L	U	100–115	D	Steam acc.
Sterilization	L	U	110	D	Steam acc.
Textile					
Dye becks	Sm	U	90–120	D	Steam acc.
Clay and Ceramics					
Waste heat from periodic kilns	L	Sm	250–350	S	U

Key: Sm = small, L = large, U = unknown, S = variable supply, D = variable demand, Steam acc. = sliding-pressure steam accumulator, Hot water = = atmospheric hot water tanks.

There are several discontinuous processes which utilize hot water or steam at low temperatures (up to about 120°C) in the food industry. Hot water or steam storage is applied.

In the textile industry, bleaching and dying processes are still discontinuous in many instances and thus require high peak loads.

Finally we note that waste heat recovery in periodically operating furnaces in the ceramic industry is a potential field for the application of TES.

5.2. Solid Storage

5.2.1. Regenerators

Regenerators (that is, heat exchangers with heat storage in a solid matrix through which two alternating gas streams are passed) are being used for preheating the air for combustion by means of hot flue gas in open hearth furnaces, in glass furnaces, in coke oven batteries, and in blast furnace stoves [5.3, 5.4]. If the solid body is stationary and the two gas streams are switched from time to time, there will be at least two regenerators, one in the heating and one in the cooling mode. If there is a relative movement between the solid body and the two gas streams, then no switching is necessary.

The main advantage of the regenerator is that it allows to achieve very high air temperature with inexpensive ceramic materials–such as magnesia (MgO), alumina (Al_2O_3), silica (SiO_2) or cast iron–and without any serious problems as regards sealing and thermal expansion. The geometry of the matrix is usually quite simple. Fig. 5.1 shows modern checkerwork arrangements. In glass furnaces, the burden from carryover, condensate and wear on the checkerwork must be taken into account.

A chimney-type checker block was developed (A) which is suitable for magnesia or chromite [5.5]. The specific surface ranges from 12 to 16 m^2/m^3, void volume from 74 to 60% and passage width from 170 to 140 mm.

Much smaller passage widths are possible in the checkerworks of hot blast stoves (B). In fact, these may be as low as 31 to 34 mm which then leads to a specific surface of from 41 to 45 m^2/m^3 and a void volume of from 32 to 38%.

The brick with a large number of triangularly arranged round gas paths (C) is a new development for regenerators of coke ovens. It has a specific surface of approximately 100 m^2/m^3 and a void volume of 47% [5.6]. Also randomly packed pebble beds offer excellent heat storage and thermal properties. It would seem, however, that stationary beds with high temperature swings are exposed to unduly high cyclic thermal expansion, situations which may cause failures of the lining or the pebbles [2.15].

A modern cowper for a blast furnace is shown in Fig. 5.2 [5.7]. Its internal volume is filled with bricks arranged in a checkerwork or similar pattern. Moreover, there is a separate furnace chamber in order to ensure the high air temperatures (up to 1350°C) that are necessary in some cases. Such an arrangement is, therefore, a combination of regenerator and furnace. Flue gas temperatures are

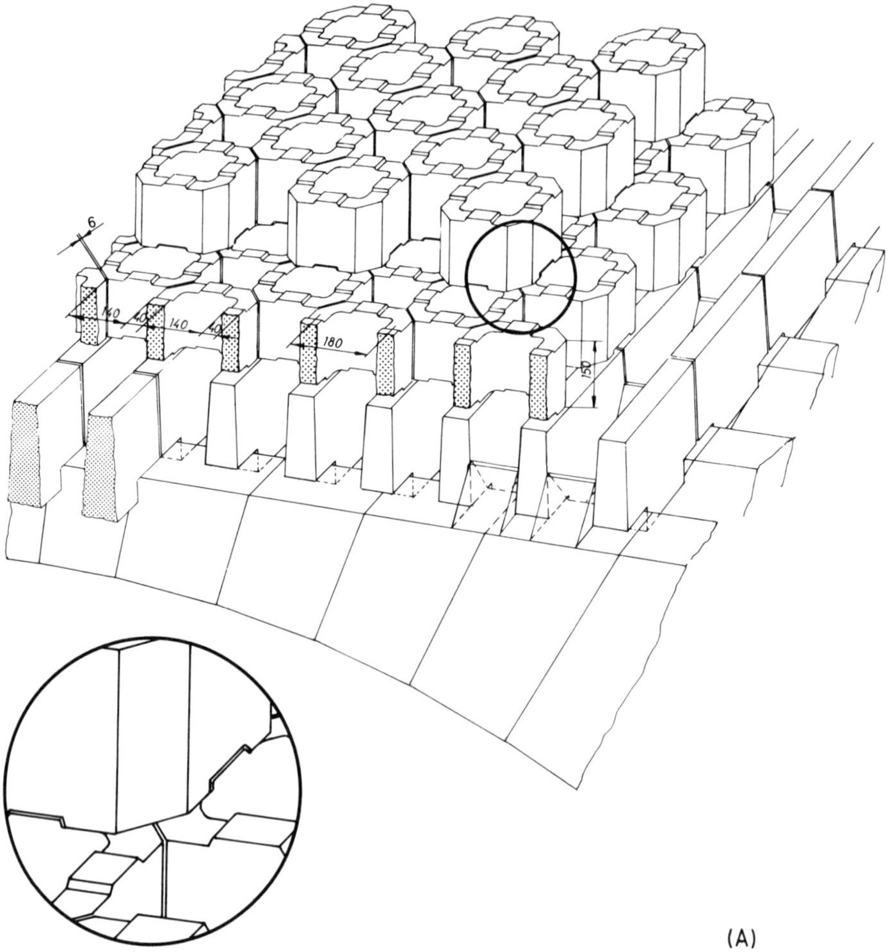

(A)

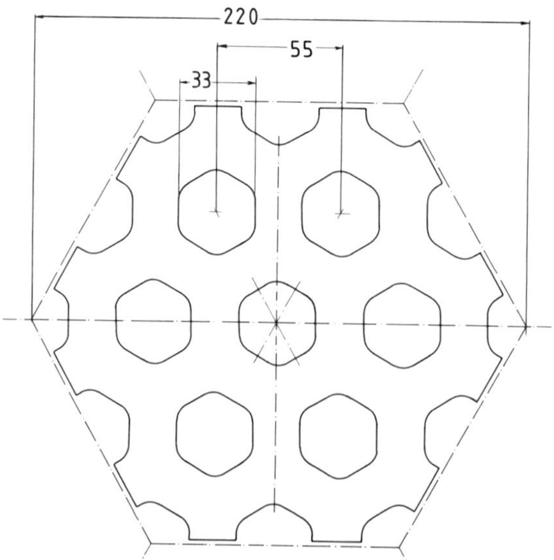

(B)

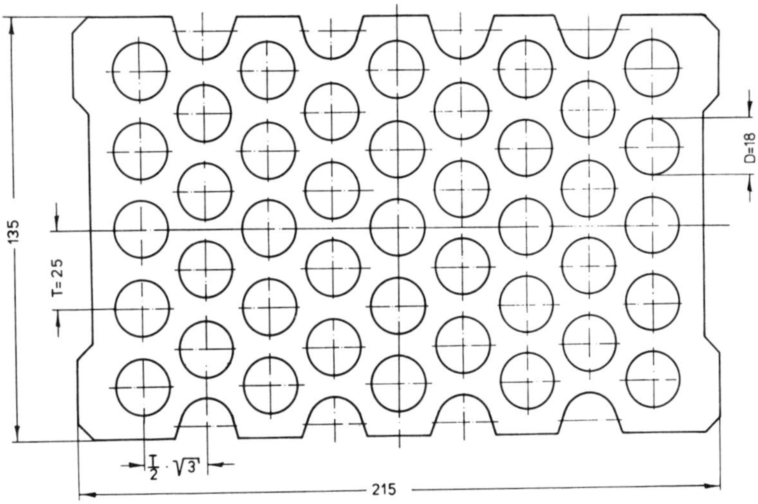

(C)

Fig. 5.1. Checkerwork arrangements–dimensions in mm
A chimney-type checker blocks for glass tank regenerators (Veitscher Magnesit-Werke AG, Austria [5.5]), *B* hot blast stove checker packing (Krupp-Koppers GmbH, Essen), *C* brick with holes for coke oven regenerators (Krupp-Koppers GmbH, Essen [5.6])

of the order of 200 to 300°C; the switching cycle is about 1.5 hours. The cowper is pressurized.

The regenerator is an important part of the plant in coke oven batteries. It is used in preheating air for combustion and, in the case of lean gas, it is used for preheating the gas. There is a pair of regenerators below the coking chambers one of which is in the preheating mode and the other one in the flue gas cooling (down to 250–350°C) mode of operation. The switching cycle is 20 minutes.

The supply temperatures of regenerators with gas stream switching are transients (that is, they decrease over the switching period). The supply temperatures of rotating regenerators are time-independent, but they show a certain local variation, the supply temperature at the inlet (in the sense of rotation) being highest. Examples of rotary regenerators are air preheaters in air-conditioning plants and air preheaters of steam generators (Fig. 5.3) [5.4] which consist of thin (less than 1 mm) corrugated steel sheets with temperature oscillations of 5 to 20 K and pinch point temperature differences (at the hot end) of less than 25 K. There is always some leakage between the two gas streams in continuously operating (rotating) arrangements. This is not of major significance in air preheaters for boilers; it is more serious in gas turbines because of the pressure difference. However, ceramic high temperature rotary regenerators are being used for open-cycle gas turbines of vehicles in which weight and space are at a premium.

5.2.2. Heat Recovery from Hot Waste Gas

Applications of solid storage other than for air preheating include the

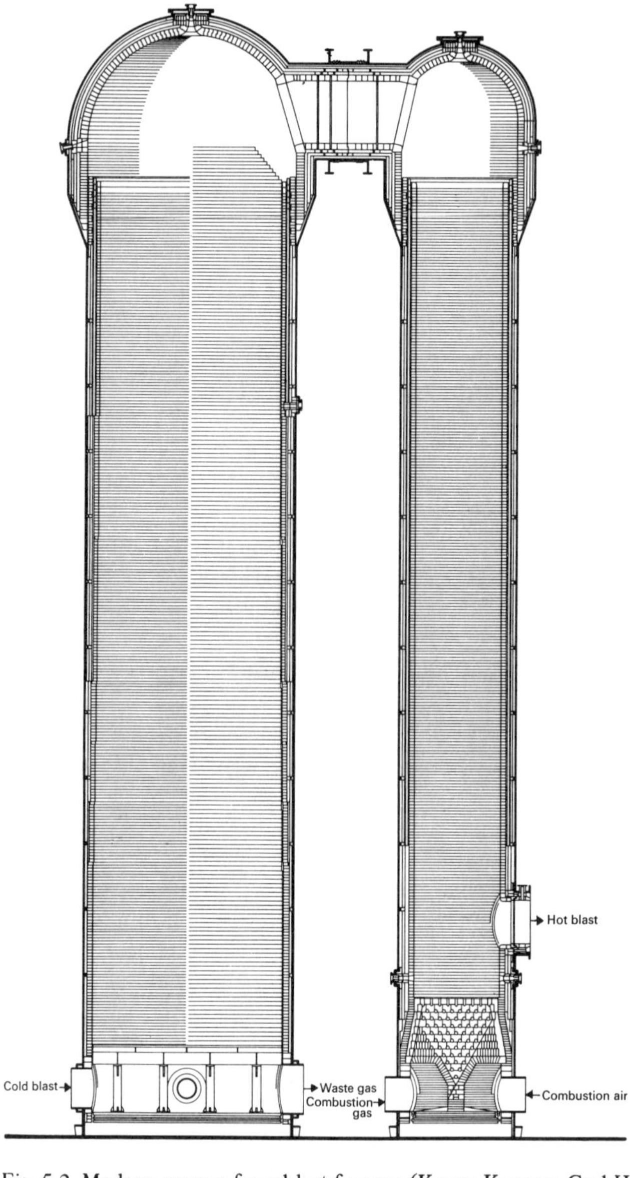

Fig. 5.2. Modern cowper for a blast furnace (Krupp-Koppers GmbH, Essen [5.7])

utilization of waste heat from intermittent streams of hot waste gas. Fig. 5.4 shows an energy recovery system for electric arc furnaces with steam generation [5.8]. The intermittent waste gas stream (which contains a sensible heat of 800 MJ/tonne steel [5.9]) must usually be cooled by a water spray before it can be cleaned. Waste heat recovery requires energy storage. In the flow sheet, two solid storage units are shown, one of them ("operational store") equalizes the

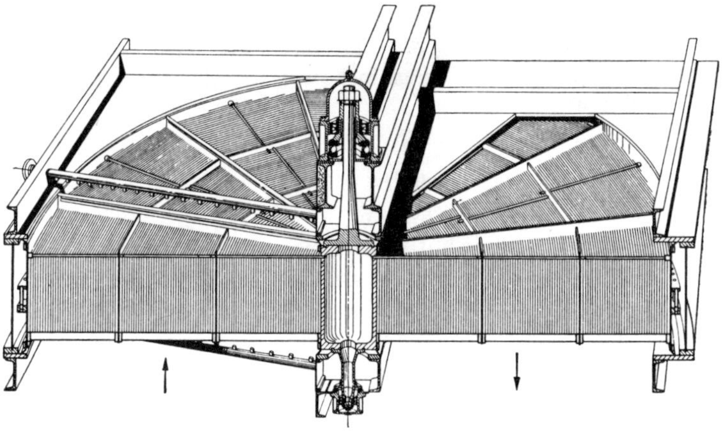

Fig. 5.3. Rotary "Ljungström" regenerator for steam generators [5.4]

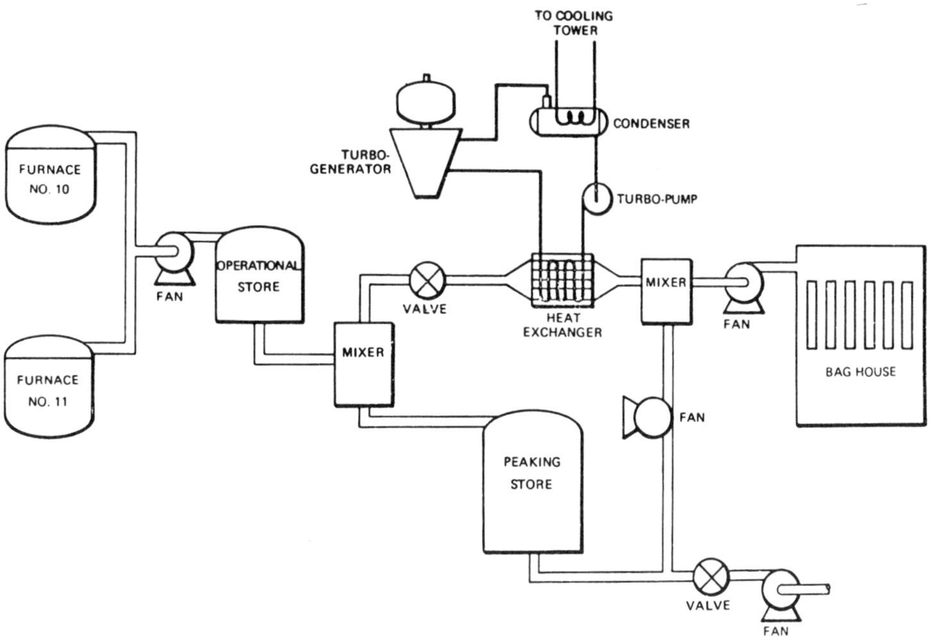

Fig. 5.4. Steel arc furnace energy recovery system [5.8]

gas temperature and the other is arranged in parallel with the waste heat boiler and is either in the charging mode (waste gas cooling) or in the discharge mode (air preheating). Steam generation is almost constant, thus enabling its use for power generation. Another possibility here is the use of sliding-pressure accumulators as it is practized in waste heat boilers behind basic oxygen furnaces (see Section 5.3.2).

Apart from steam generation, the heat from the waste gas can be directly used for the preheating of the incoming raw material which itself acts as a heat storage medium. Preheating leads to savings in power and electrodes as well as to corresponding increase of throughput.

5.2.3. Heat Recovery from Hot Solid Products

Utilization of the sensible heat of hot solids which comes from a periodically operating process (kiln, retort, oven, etc.) is another field for industrial TES (especially for solid storage).

Although drying and preheating of the incoming material for periodic kilns in the brick and ceramic industries by storing and utilizing the sensible heat of the hot product may be interesting from the standpoint of energy saving, in many cases it may be uneconomical because of the long time cycle (up to several days) involved.

Dry cooling of coke (DCC) recovers the sensible heat of incandescent coke economically [5.10, 5.11]. Coal is coked in horizontal retorts (ovens); at the end of the coking time (16 to 20 hours), the coke is discharged ("pushed") from the oven at approximately 1000°C. Because a coke oven battery consists of a large number of ovens, the pushing interval is generally from 10 to 20 minutes. However, longer pushing intermissions (30 to 90 minutes) will occur during the change of a working shift. If the recovered heat is used for steam generation, the yield amounts to about 0.4 tonnes of steam per tonne of coke, depending on steam and feed water parameters.

The method of dry coke cooling has a tradition of more than 50 years. It was used by Sulzer Brothers (Winterthur, Switzerland) in both Europe and the United States at more than 70 steel mills and gas works. Waagner-Biro AG now owns the rights for this Sulzer DCC system. Cooperation between Krupp Koppers GmbH and Waagner-Biro AG has spurred further development of the DCC process, leading to the Encoke process which is shown in Fig. 5.5 [5.11]. The incandescent coke is pushed out of the oven and into a transfer bucket. The full transfer bucket is then transported to the hot coke entrance of the cooling shaft. By opening the cooling-shaft cover and the bucket bottom gate, the coke dumps into the cooling shaft without any dust being allowed to escape.

The cooling shaft is provided with inserts for mass flow. The coke passes through the cooling shaft over a period of about 2 to 4 hours. While the coke is cooled to 200°C and discharged through an air lock, the counterflow of cooling gas is being heated from 130°C to more than 700°C. As it leaves the cooling shaft, the hot gas passes into a coarse-particle separator which reduces the dust load on the waste heat boiler. The boiler is a water tube boiler with an economizer, an evaporator and a superheater. The cooling gas passes through a fine-particle separator and is recirculated to the cooling shaft. TES in the DCC process allows high grade steam to be generated with almost no fluctuations in steam parameters. This is so even during pushing intermissions. The hot coke itself serves as a heat storage medium. Two basic principles of heat storage are possible and feasible:

(a) *The Single-chamber Process*

In this process, the cooling shaft is a single structure which is not subdivided. Storage and cooling take place in the same chamber. All Sulzer-DCC plants worked on this principle. Heat storage capability can be activated by increasing the gas flow through the cooling shaft. The resulting fluctuations in hot gas outlet temperature can be eliminated by a by-pass line (see No. 15 in Fig. 5.5) and/or by a dual gas feed at the lower end of the cooling shaft (see No. 14 in Fig. 5.5).

By controlling the by-pass flow rate and mixing the by-pass gas with the hot gas from the cooling shaft, the gas temperature at the boiler inlet (and thus the steam generation) may be controlled. The whole coke content acts as a sliding-temperature sensible heat storage. Fig. 5.6 shows the coke temperature profile

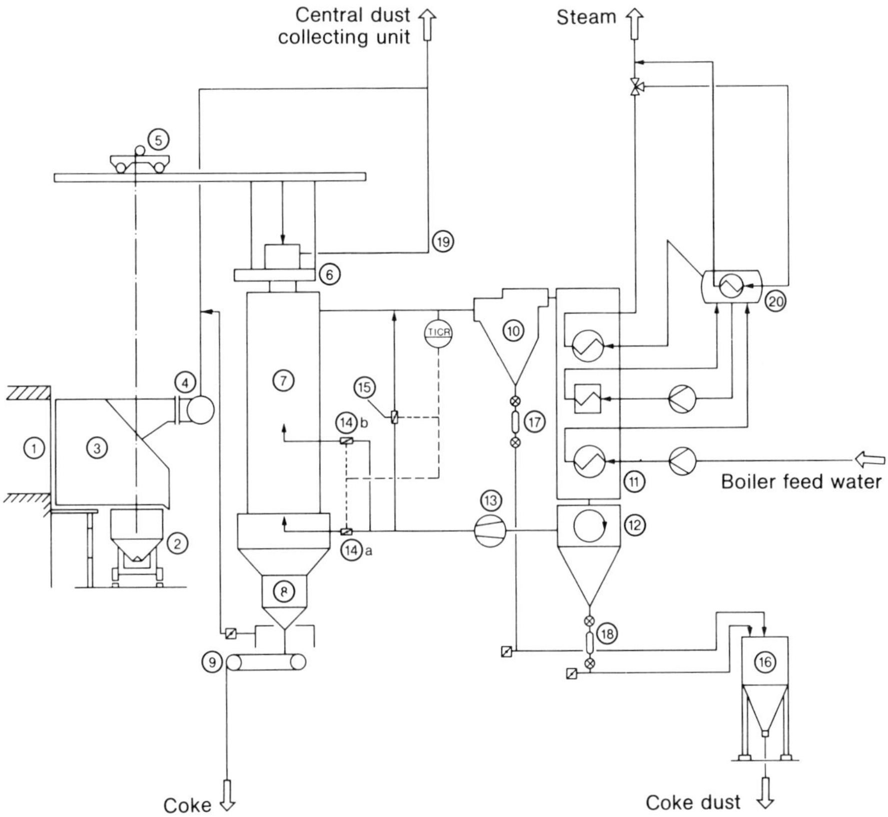

Fig. 5.5. The Encoke-process for dry coke cooling (DCC) [5.11]
1 coke oven battery, *2* coke car, *3* coke guide car with dust-collecting hood, *4* suction line, *5* crane, *6* hot coke entrance, *7* cooling shaft, *8* cooled-coke extractor, *9* conveyor, *10* coarse-coke separator, *11* boiler, *12* coke fines separator, *13* recycle gas blower, *14* recycle gas feed in two stages, *15* by-pass, *16* coke dust hopper, *17* coke dust sluice with cooling system, *18* coke dust sluice without cooling system, *19* smokeless charging unit, *20* boiler drum

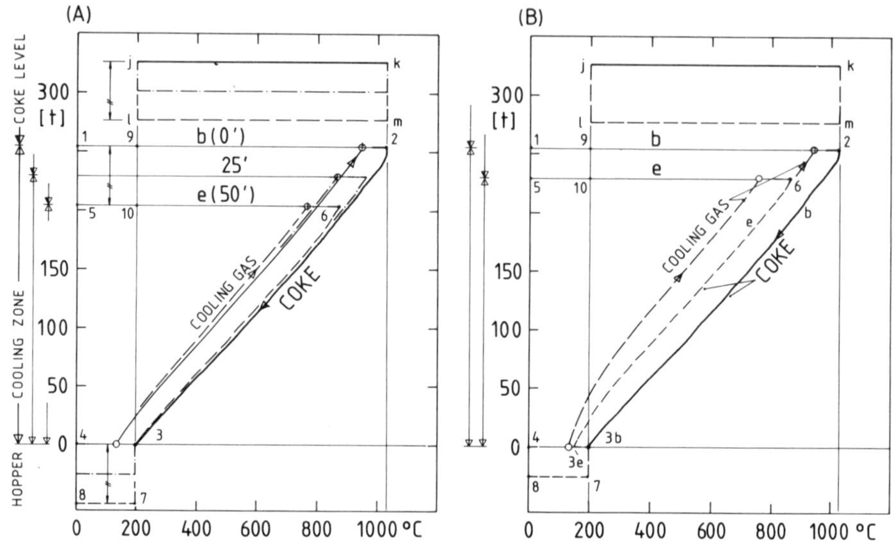

Fig. 5.6. Coke and gas temperature profile in the cooling shaft of a single chamber DCC unit with a coke throughput of 60 t/h
A continuous coke discharge during coke pushing intermission, *B* stopped coke discharge in the middle of the coke pushing intermission
b beginning of coke pushing intermission, *e* end of the 50 minute coke pushing intermission

and the corresponding gas temperature of a DCC unit with a coke throughput of 60 tonnes/h during a 50 minutes coke pushing intermission (discharge mode), in which constant steam generation is maintained (the amount of heat which is discharged during intermission equals the difference between the heat content in the cooling zone at the beginning and the heat content of the same coke mass at the end of the intermission). The picture A at the left shows the temperatures during a continuous cooled coke discharge, the picture B on the right shows the conditions if the cooled coke discharge is stopped in the middle of the intermission. (Heat content of the coke at the beginning of the intermission is proportional to the area 1–2–3–4. Heat content at the end is represented by the area 5–6–7–8. The difference (and hence the amount of heat discharged) is represented by the area 9–2–3–6–10 or *j*–*k*–*l*–*m*.) As can be seen, heat storage capability is nearly independent of the cooled coke discharge rate.

(b) *The Two-chamber Process*

This process is also based on Sulzer patents, but it has not been applied by Sulzer. In this process, the incandescent coke is initially stored in the prechamber without being cooled. The hot coke passes into the actual cooling chamber only by the way of the coke that is withdrawn from the cooling shaft (displacement principle). The hot gas has to be withdrawn from the cooling shaft between the prechamber and the cooling chamber. In the Soviet Giprokoks process this is accomplished by a number of slots and a concentric annular collecting duct.

Lining and refractories in the cooling bunker are rather sophisticated. The steam rate is not controlled but rather depends on the cooled coke discharge rate. (If the coke discharge is interrupted, then steam generation drops to approximately 75% within 10 minutes.)

When compared to the original Sulzer process (a process without active heat storage), modern DCC processes represent a substantial improvement in terms of heat storage capability and hence in terms of flexibility and efficiency of heat recovery also.

5.3. Hot Water Storage

5.3.1. Jet Pumps

Contrary to solids, hot pressurized water is not only suitable for TES but also for providing a propellant for pumps. Intermittent operation of jet pumps may be performed by means of sliding-pressure storage, with hot water or steam discharge.

Air flow test rigs (in particular for supersonic flow) require high blower power. Intermittent operation, with several minutes operating time, is often sufficient to achieve steady-state flow conditions. Hot water storage will economically supply the motive power in such instances. Fig. 5.7 shows such an arrangement at the German Aerospace Institute (DVFLR) in Brunswick, Federal Republic of Germany. The test rig operates with Mach numbers of 2 to 3.3 in a quadratic cross-section of 0.7 meters. Two horizontal storage vessels of the expansion type (100 m^3 each, 100 bar) supply hot water for a period of 7 minutes to the ejector of 50 MW power. After the ejector, there is a water separator which minimizes water input to the cycle. Charging is performed indirectly via a heat exchanger [2.7a].

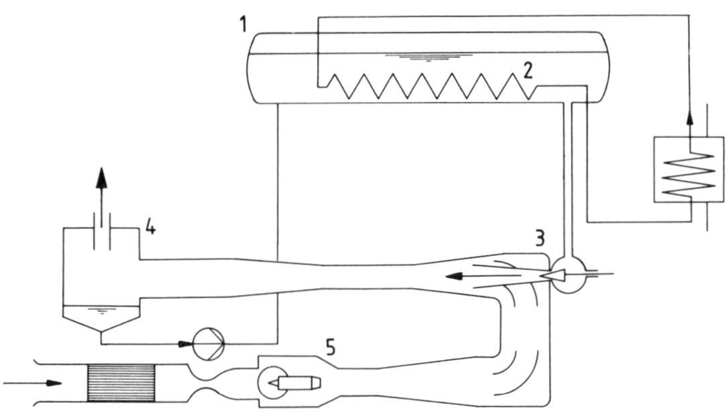

Fig. 5.7. Flow sheet of a supersonic air flow test rig [2.7a]
1 expansion-type store, *2* heating coil, *3* ejector, *4* water separator, *5* test channel

Waste gas is produced intermittently in many industrial processes. If the operating periods of the waste gas cleaning plants are short, a wet scrubber may be economically combined with a steam jet pump rather than with an induced draft fan, the steam being supplied by a steam storage installation. As an example, Fig. 5.8 shows a mobile wet scrubber for pushing emission control in a coking plant. The sliding-pressure store operates at between 23 and 10 bar, and coke is pushed every 6 minutes for about 1 minute. Although the coke trolley is supplied with electric power via a moving cable, the steam storage system was installed because of the high peak power requirement [5.12].

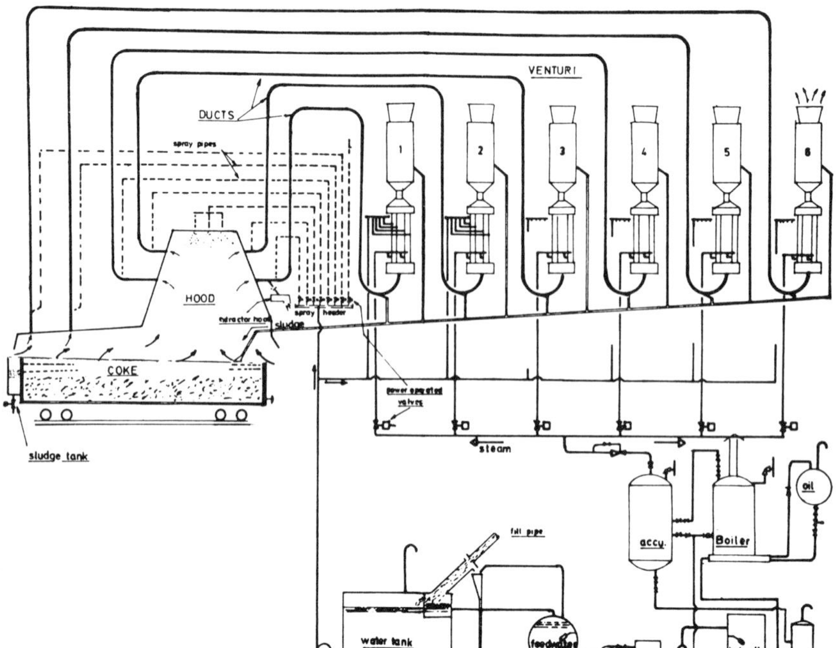

Fig. 5.8. Coke-pushing emission control system with mobile Venturi scrubber and steam accumulators [5.12]

5.3.2. Storage Integrated with a Waste Heat Boiler

If constant steam or energy output is required, then waste heat boilers behind periodic operating processes can be integrated with a hot water store. Waste heat boilers behind basic oxygen furnaces (BOF) in steel mills are typical examples of this.

Blowing times of 15 to 20 minutes with high peaks of up to 1600°C and steep gradients of waste gas temperature, intervals of 20 to 30 minutes and total periods of 35 to 50 minutes are typical for such plants. Exhaust gas from BOFs contains per tonne steel approximately 100 MJ sensible heat and an additional 500 MJ calorific value due to the CO content of 70% [5.9]. Combustion of CO gas may be either complete (which gives the largest heat load at the hood) or suppressed. This latter arrangement results in the hood with the smallest heat load, and the cooled gas has to be stored as gaseous fuel. Even if two converters are operated alternately and the resulting short intervals are bridged by conventional firing, steam storage is still useful. Fig. 5.9 shows a BOF plant with complete combustion. There are three converters, two of which are in operation, while the third is held in reserve. The boiler operates at constant pressure which is much higher than that of the steam grid. Between the boiler live steam line and the steam grid a sliding-pressure accumulator is arranged. This is charged via a pressure control valve and discharged via a pressure reduction valve. A fossil-fired superheater may be arranged before the steam turbine, generating driving power or electricity. Many plants of this type have been built all over the world [5.13].

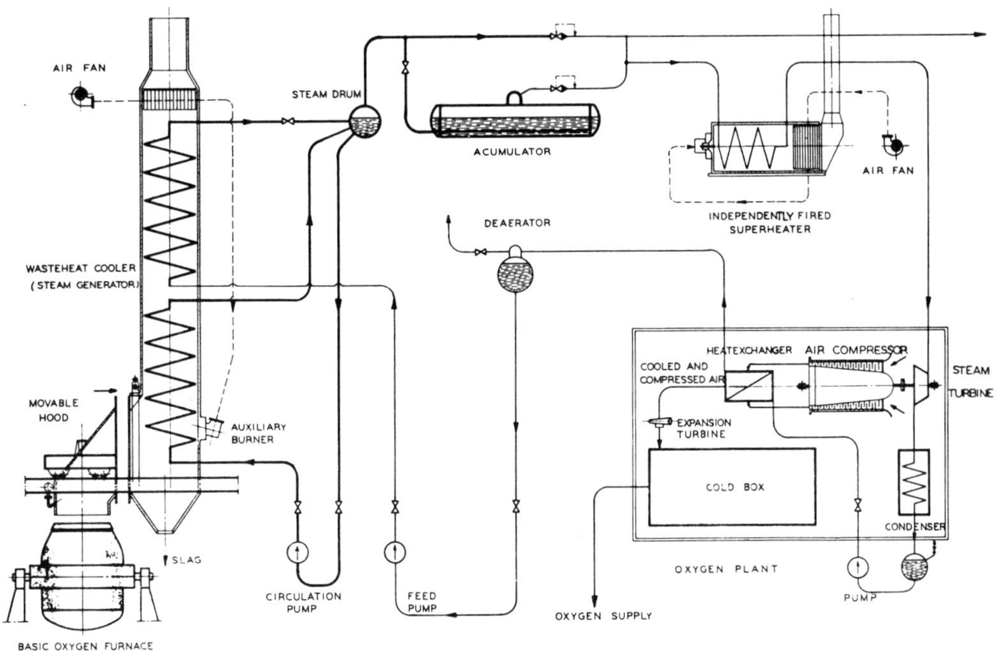

Fig. 5.9. Flow sheet of a BOF plant with a steam accumulator [5.16]

The alternative scheme of waste heat recovery and storage in a Japanese BOF steel mill [5.14] is shown in Fig. 5.10. Water is heated to 98°C in the hood of the waste heat boiler and is fed into the unpressurized higher temperature tank and stored. The hot water transfers continuously the heat to an organic Rankine cycle based on Freon-11 as a working fluid. Freon vapor (4.6 bar, 75°C) drives the turbine, is condensed, and the condensate is recirculated [5.14]. The

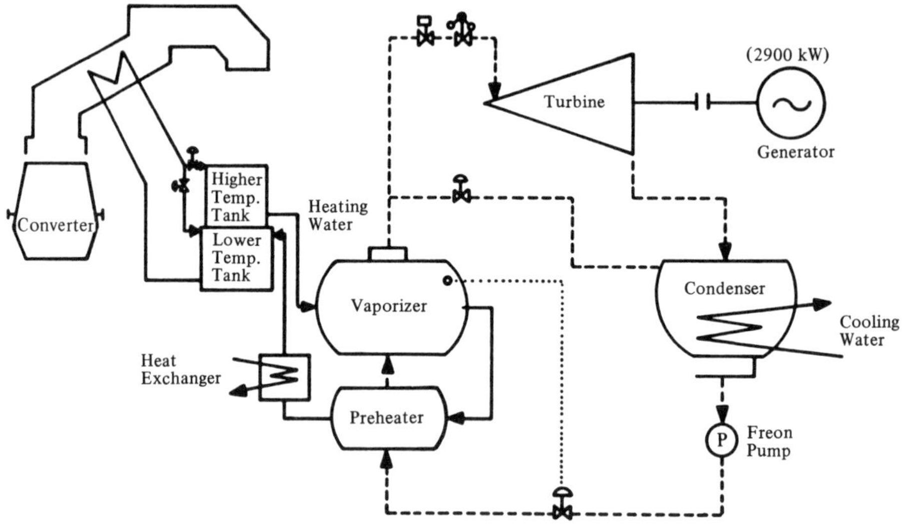

Fig. 5.10. Flow sheet of a BOF plant with displacement accumulator and organic Rankine cycle [5.14]

output of the turbine strongly depends on the temperature of the cooling water. It is 2.9 MW during winter conditions (cooling water temperature 8°C) and only 1.4 MW during summer conditions (cooling water temperature 29°C) [5.15]. The thermal efficiency is below 9% which is less than one third compared with the previously mentioned scheme. On the other hand, however, boiler and storage system are operated at a low pressure level.

5.3.3. Storage in Industrial Steam and Hot Water Grids

Steam storage is utilized in industrial steam grids if variations of steam demand or supply are to be equalized. The pressure in industrial steam grids may be fixed or some variation may be permitted as in most low pressure grids [5.1, 5.2]. In the latter case, steam storage may be directly connected to the grid via two non-return valves (Fig. 5.11 A). Storage capacity is limited by the allowable pressure swing in the grid. In most cases, there will be at least two grids with different pressures between which the steam store may be arranged (Fig. 5.11 B). Here, too, there are non-return valves. A pressure-reducing valve is actuated by the pressure in the upper grid (IP), keeping it (within the limits of the storage capacity and the requirements of the lower grid) at a constant level without an elaborate control system. If the IP steam is superheated and superheating of LP steam is required, a superheat store (shown by dashed lines) is to be installed. Fig. 5.11 C shows an alternative arrangement. The IP line is connected via a pressure control valve to the charge pipe, while the LP line is connected via a pressure reduction valve to the discharge pipe. No non-return valves are necessary, but discharge and charge pipe now have to be designed for the total flow rate

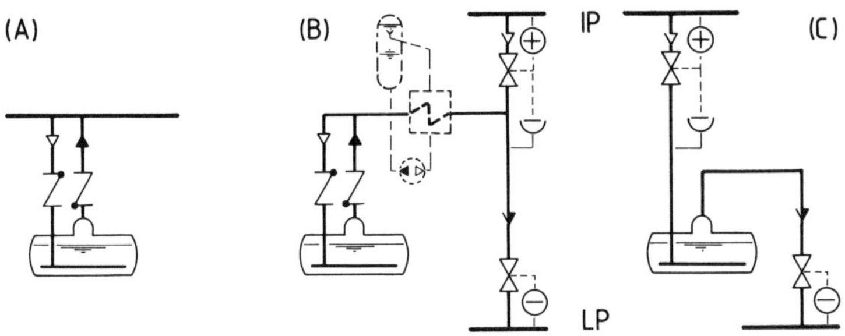

Fig. 5.11. Arrangement of steam accumulators in steam grids
A sliding pressure grid, *B* steam accumulator in the side stream of IP and LP grid, *C* steam accumulator in the main stream between IP and LP grid

through the pressure control and pressure reduction valve, whereas in a scheme according to Fig. 5.11 B, only the difference between these two flow rates determines the design.

In a similar manner, demand peaks may be covered by steam storage. Pulp and paper, food and textile are typical industrial branches with such demand peaks. The generally low pressures that are required in these branches are an advantage because the pressure swing between charge and discharge (and therefore the specific storage capacity) may be high. The savings stemming from steam storage are given by the reduction of boiler power and the minimization of its load transients.

Combined heat and power generation (CHP) is another industrial application of steam storage. Fig. 5.12 A presents a flow sheet with steam turbine between a constant pressure HP and a variable pressure LP steam grid with steam accumulator which permits maximum electric power generation in that all the LP steam that is required is passed through the turbine which is designed for the average steam rate only.

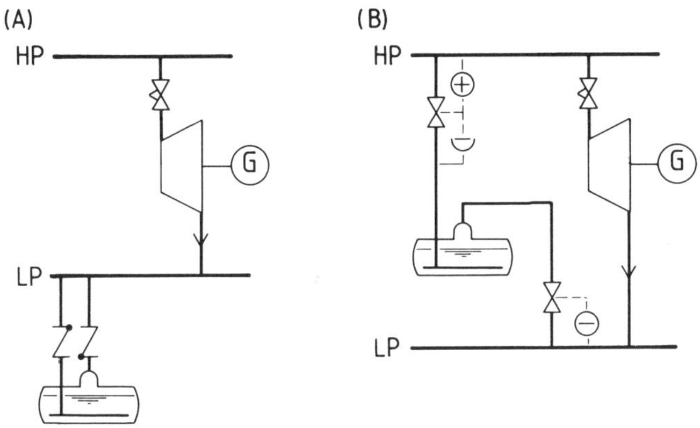

Fig. 5.12. Combined heat and power (CHP) plant
A steam accumulator behind the turbine, *B* steam accumulator parallel to the turbine

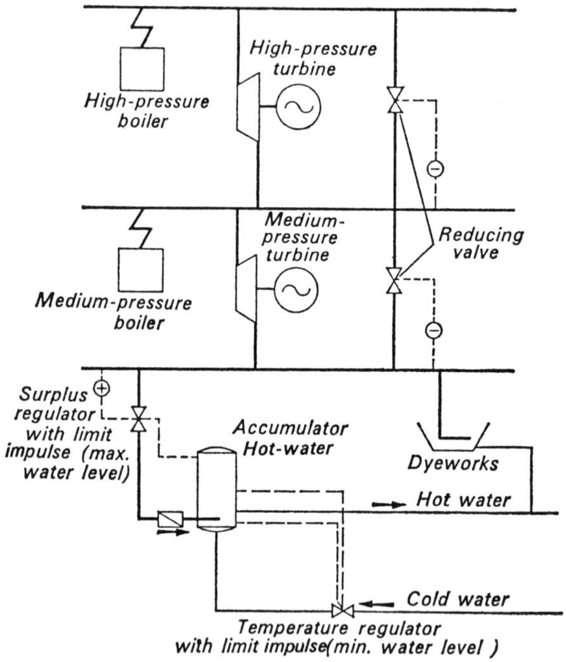

Fig. 5.13. Hot water accumulator in a carpet factory [5.2]

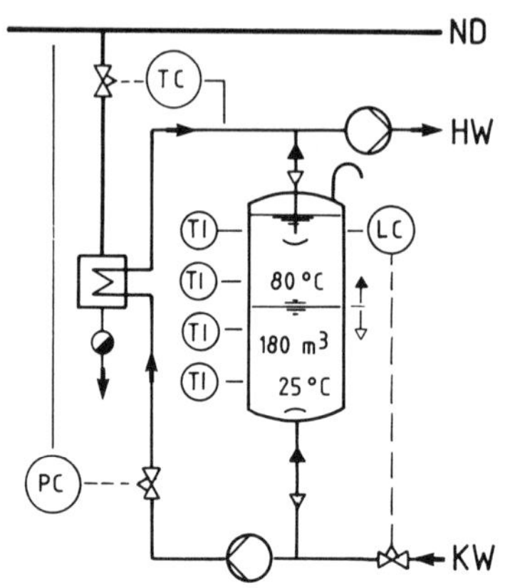

Fig. 5.14. Displacement accumulator in a molasse utilization factory
ND low pressure steam, *HW* hot water supply, *KW* cold water supply

In the flow sheet according to Fig. 5.12 B, the steam accumulator is connected in parallel to the turbine between the HP and LP grid. If the turbine is to operate at base load, then only the minimum LP steam passes through the turbine, while the remainder passes through the store.

If there is a hot water grid–as in a textile factory having a dyeing plant (Fig. 5.13)–the constant pressure hot water store may be connected to the fixed pressure LP steam grid. Again, all the LP steam is expanded in the turbine [5.2].

In pure hot water grids, mainly displacement storage is used. Fig. 5.14 shows the hot water supply for a molasse plant. Hot water is produced by LP steam, stored in a vertical displacement storage vessel and distributed by supply pumps to the consumers. Peaks of hot water demand are covered by the store. In addition, peaks of LP steam demand (for sterilizing the fermenters) are balanced by closing the charge valve (this mode of operation is called indirect balance).

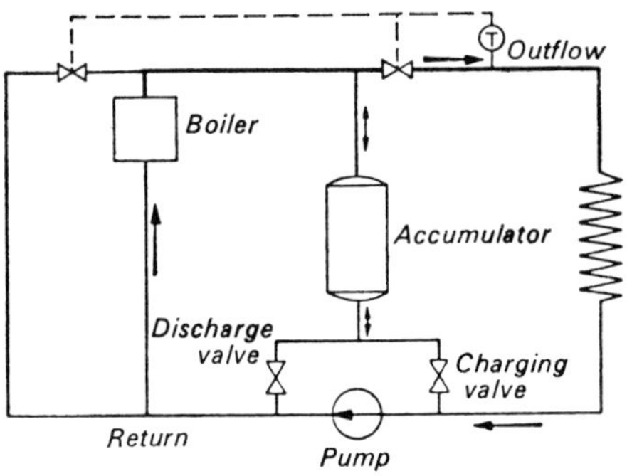

Fig. 5.15. Displacement accumulator in an industrial heating plant [5.2]

Fig. 5.15 shows the flow sheet of a heating plant, with displacement storage between supply and return line. By opening the discharge valve and closing the charge valve, the major part of return water is directed into the store and the same volume of hot water is discharged to the supply line. In the charge mode, part of the supply water from the boiler is directed to the store. The recirculation pump is to be designed for the maximum sum of demand plus charging. For high charge power, therefore, pumping power is rather high and a flow sheet with separate charge pump as used in district heating plants (Section 6.6) may be preferable. If the boiler (as is common in small plants) is equipped with on-off control, considerable fuel savings are possible, in particular at part load, because of a reduced frequency of operation (see Section 6.2).

5.4. Conclusions

Mainly short-term solid storage for regenerators as parts of industrial furnaces and short- to medium-term hot water storage (pressurized and unpressurized) for heat and/or power storage is used in industrial plants. Storage in cogeneration plants receives increasing attention [1.16]. This is because of the growing portion of recovered waste heat which is available in excess of the instantaneous demand.

6. Thermal Energy Storage for Space Heating and Cooling

6.1. General

Heat generation in heating systems is usually controlled according to the instantaneous demand; the design power of the heating system is chosen according to the maximum demand; and energy–for fuelled heating plants–is stored as fuel before its conversion to heat. In the future, storage after energy conversion, that is, storage of heat (secondary energy), will become more important in space heating and in hot tap water production. And this is so for the following reasons:

- The design power of energy conversion plants and of secondary energy transportation systems (district heating piping and network) may be reduced by means of TES, since maximum heating power is required for only a short portion of the year, that is, for a few days or even for only a few hours.
- In many cases, the TES facility may act as a reserve unit.
- The efficiency of energy conversion may be improved by TES (no part-load operation; reduction of start-up and shut-down losses by buffer storage).
- In electric space heating and hot tap water production, favourable rates may be obtained by shifting the electricity consumption to off-peak hours.
- For solar heating and hot tap water production, the solar coverage may be considerably improved by short-term storage (hours, days and periods of up to one week) and by long-term storage (weeks and months, up to seasonal storage).
- The coefficient of performance (COP) as well as its average value–the seasonal performance factor (SPF)–of heat pumps with ambient air (or with solar radiation) as the heat source may be improved by short-term storage (avoiding operation at minimum ambient air temperature and maximum heating demand during the early morning hours) and by long-term storage also.
- Thermal storage may decouple the two products of combined heat and power (CHP; cogeneration) plants according to the demand for each of the two products. The demand for these varies according to several periodicity patterns:
  diurnal (ambient air temperature, solar radiation, consumer habits);
  several days (typical weather period);
  weekly (low industrial and commercial electricity demand during weekends);

seasonal (winter heating demand, summer air-conditioning demand, more or less constant hot tap water demand).

Aspects of the applications of short-term storage to heating boilers, electric heating, solar heat, heat pumps and cogeneration are dealt with first. This discussion is then followed by some common considerations that are relevant to large-scale long-term storage.

6.2. Heating Boilers

Oil- and gas-fired automatic heating boilers for single-family homes and for medium to large houses are presently equipped with on-off control. The frequency of boiler operation depends not only on the ratio of instantaneous power demand to boiler plant power, but also on the difference between the on and off temperatures and on the system storage capacity. The higher the frequency of boiler operation, the higher will be the start-up and shut-down losses in absolute terms and, even more so, relative to the heating energy in the time period considered.

Short-term storage (buffer storage) as shown in Fig. 5.15 will improve matters by increasing effective system storage capacity and will thus considerably reduce the frequency of boiler operation. Buffer storage will also reduce fouling of the heat transfer surface which depends on the number of on-off operations. High frequencies of boiler operation may also have detrimental effects on the stack (which may never reach its minimum working temperature at part load operation) such as prolonged periods of operation below dew point, stack wetting and reduced draft. For extreme part loads (for example, in the case of boilers for space heating and hot tap water that are operated in summer for hot tap water only) buffer storage is even more important. The low boiler efficiency quoted for this mode of operation may be considerably improved by buffer storage. However, the gain will be somewhat reduced by storage losses. Therefore, these should be kept to low values by proper thermal insulation.

Conditions are similar with boilers for solid fuels (coal, coke, wood, waste). There may be some load control capabilities by means of draft control. However, in most cases the main control is an on-off control with prolonged operating period. Buffer storage (for about 50 to 100% of maximum daily demand) will be of particular advantage for fuels with fast combustion such as wood, since it will increase efficiency and simplify operation.

Apart from reducing start-up and shut-down losses by thermal storage, there is a further saving of energy if the period in which the boiler is kept at operating temperature can be reduced. This will be possible in summer when the boiler is only used for hot tap water production, as well as during the heating season when boiler operation is interrupted (for instance, during night time or during weekends).

Assuming a conventional boiler, maintained at constant operating temperature with on-off control of the burner, and defining

Z_o	time in which the boiler is kept at operating temperature
Z_f	working hours of the burner
Z_u	useful working hours of the burner
$Z_s = Z_f - Z_u$	working hours of the burner to compensate for (surface and stack) heat losses outside its useful working time
$f = Z_f/Z_o$	load factor of burner
$q = Z_s/(Z_o - Z_u)$	stop loss factor
η_B	boiler efficiency
η_{Bs}	seasonal efficiency
$f_s = \eta_{Bs}/\eta_B = Z_u/Z_f$	part load factor

the stop loss factor may be written as [6.2]

$$q = Z_s/(Z_o - Z_u) = (Z_f - Z_u)/(Z_o - Z_u) = (1 - Z_u/Z_f)/(Z_o/Z_f - Z_u/Z_f). \quad (6.1)$$

And after some arithmetic, we have

$$f_s = \frac{Z_u}{Z_f} = \frac{1 - q/f}{1 - q}. \quad (6.2)$$

The seasonal efficiency becomes

$$\eta_{Bs} = f_s \eta_B = \frac{1 - q/f}{1 - q}\,\eta_B. \quad (6.3)$$

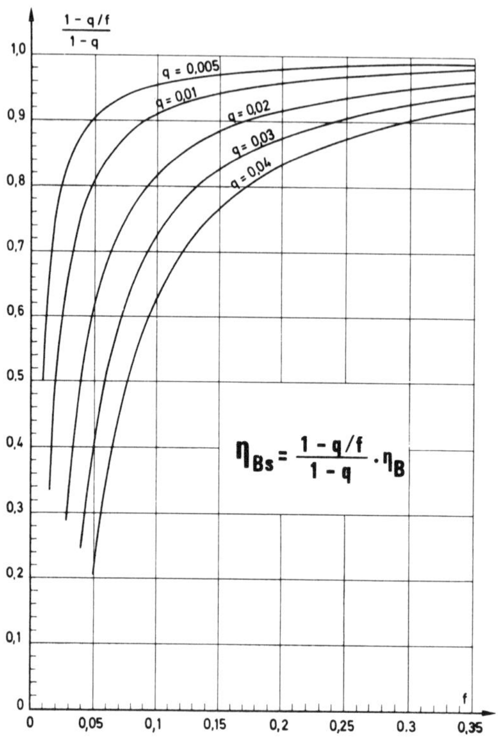

Fig. 6.1. Part load factor $f_s = (1 - q/f)/(1 - q)$ of a boiler as a function of load factor of burner f [6.2]

In Fig. 6.1 the part load factor f_s–a measure for the seasonal efficiency at given (full load) boiler efficiency–is shown as function of f, the burner load factor, and q, the stop loss factor. For instance, if for all-year operation $f = 0.15$ and $q = 0.03$, then $f_s = 0.8/0.97 = 0.825$. Thermal storage of suitable size may reduce Z_O from 8760 to, say, 5000 h/a and, therefore, increase f to 0.26 and f_s to $0.885/0.97 = 0.912$.

However, for summer operation only, f will reduce to, say, 0.05 and the part load factor becomes $f_s = 0.4/0.97 = 0.412$. With thermal storage, necessitating boiler operation only every few days for a relatively short time, f may increase to, say, 0.5; and, correspondingly, $f_s = 0.97$. This large improvement will, of course, be somewhat reduced by cooling-down losses and by insulation losses of the store. On the other hand, it will be increased by reduced start-up and shut-down losses of the burner.

Conventional water tanks are normally used as storage vessels in connection with heating boilers. For hot tap water, corrosion resistant material or coating may be required.

6.3. Electric Heating

Electric off-peak resistance heating with thermal storage has gained wide acceptance for space heating and, even more so, for hot tap water production in many countries [6.3].

For space heating purposes, the following systems are in use:

– units in individual rooms, consisting of bricks of ceramic refractories such as

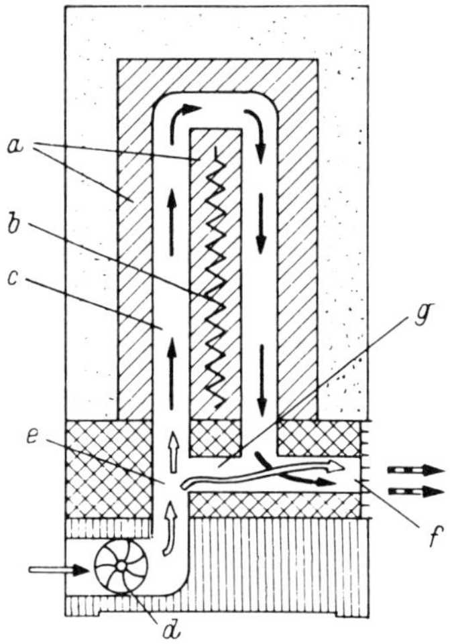

Fig. 6.2. Electric off-peak storage heater [6.3]
a storage mass (magnesia bricks), *b* heating coil, *c* air space, *d* blower, *e* air inlet, *f* air outlet, *g* bypass

magnesia with storage temperatures of about 600°C, well thermally insulated, with discharge by means of blowers that are controlled by electric thermostats (Fig. 6.2);

- electric floor heating without or with increased thickness of the floor slab to bridge the (2 to 16) peak load hours;
- central units with magnesia bricks and air as heat transport medium;
- central units with water (sometimes with cast iron also) as storage medium and having water for heat distribution.

Thermal storage of hot tap water ranges from small decentralized water stores for hand-wash basins to central hot water tanks for apartments and homes (typically charged with off-peak power) and to large units for blocks of flats and for commercial buildings.

6.4. Solar Heat

6.4.1. General

The supply of solar heat shows a regular diurnal and annual pattern and, moreover, it also shows a stochastic pattern that is given by the weather conditions. These patterns may be described by their variances, the weather pattern (whose period is not constant) as well as by its "persistence" for the particular season. For example, this is typically 3 to 4 days for Central European and most other continental climates. There are also variances in heat demand. An autocorrelation function may therefore be developed [6.4a, 6.4b].

In Table 6.1, the periodicities of solar supply and demand for heating, cooling and hot tap water are listed. There is usually a strong negative coupling between annual solar supply and space heating demand (that is, in winter solar supply is

Table 6.1. *Periodicities of Solar Heat*

Period	Supply	Demand		
		Space Heating	Space Cooling (and Dehumidification)	Hot Tap Water
(a) diurnal	regular	some periodicity		stochastic (some periodicity)
(b) weather	stochastic ("persistence")	negative coupling	positive coupling	–
(c) week	–	some periodicity		some periodicity
(d) annual	regular	strong negative coupling	strong positive coupling	–

low, heating demand is high), a strong positive coupling between annual solar supply and cooling demand (including dehumidification demand), a somewhat weaker positive coupling regarding weather between solar heat supply and cooling demand and an even weaker negative coupling regarding weather between solar heat supply and heating demand (that is, ambient air temperature during the heating season may be lower when the sky is clear—and therefore solar supply is high—than when the sky is cloudy). It follows from Table 6.1 that the role of TES in solar heat utilization is a manifold and important one.

6.4.2. Sensible Heat Storage

A typical flow sheet of an active system for hot tap water generation with sensible heat TES shows a primary cycle with anti-freeze, a heat exchanger in the lower part of the storage tank and an auxiliary heater in the upper part (Fig. 6.3). Since the efficiency of the solar collector depends strongly on the temperature difference between primary cycle and ambient, the primary cycle temperature should be kept as low as possible by a low temperature drop in the heat exchanger, by avoiding mixing in the tank and by transferring heat always to the coldest part.

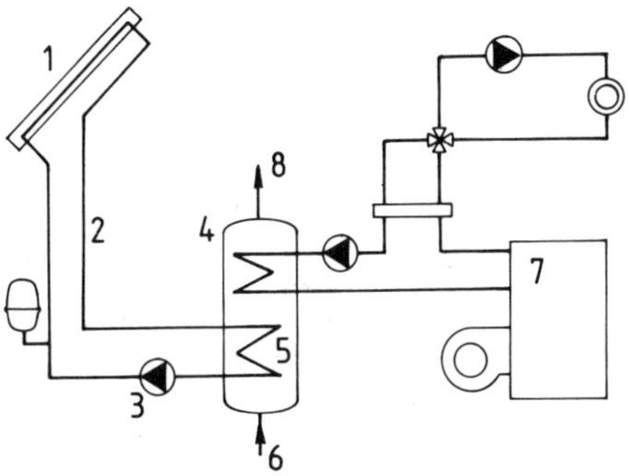

Fig. 6.3. Flow sheet of solar hot tap water production
1 solar collectors, *2* primary cycle (anti-freeze), *3* recirculation pump, *4* storage tank, *5* solar heat exchanger, *6* cold water from mains, *7* auxiliary heater, *8* supply

The trade-off between collector size and storage size for short-term (hot tap water) and long-term (solar heating) TES plants is an interesting optimization task. More correctly stated, the overall optimum is found where collector size and storage size both have their respective optima [6.8, 6.9, 6.10]. Specific storage sizes of 50–100 kg water per m^2 collector area are common for short-term storage, while sizes of the order of 1000 kg water per m^2 collector area will be necessary for long-term storage in a Central European climate.

6.4.3. The Solar Pond

The solar pond, where collector and store are combined, is a special case of

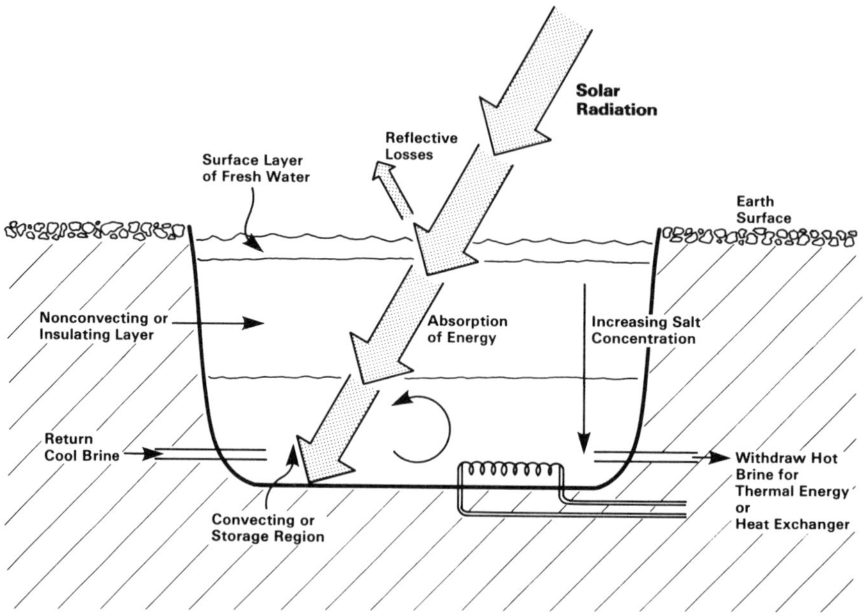

Fig. 6.4. Schematic of a salt gradient solar pond [6.11]

sensible heat storage (Fig. 6.4). Solar radiation is absorbed at the bottom of the uncovered pond. A salt density gradient (the concentration increasing with depth) is imposed and maintained between an upper convection layer (due to wind) and a lower convection layer (due to heat extraction). In this manner convection and convective heat transfer to the surface is suppressed, the non-convective layer of about one meter thickness between the two convection layers acting as thermal insulation.

Temperatures of over 100°C can be obtained in this way, with 90°C being the usual design point in hot climates [6.10].

6.4.4. Latent Heat Storage

Latent heat storage systems for charge and discharge by air (Fig. 6.5) or by water (Fig. 6.6) have also been proposed and developed [6.5–6.7]. Fig. 6.7 shows the Finned-Annulus Heat Exchanger concept with separate flow paths for the charging and the discharging fluid. The heat exchanger thus permits simultaneous charging and discharging also. Each element consists of an inner tube and an outer tube maintained in thermal contact with each other through longitudinal fins made from heat conducting material (for example, aluminum). The region in the annulus (between the fins) is filled with the latent heat (heat of fusion) storage material. The concept permits operation of the heat storage system as a hybrid latent/sensible heat store. The sensible heat is stored in the liquid surrounding the finned-annulus elements in the containment.

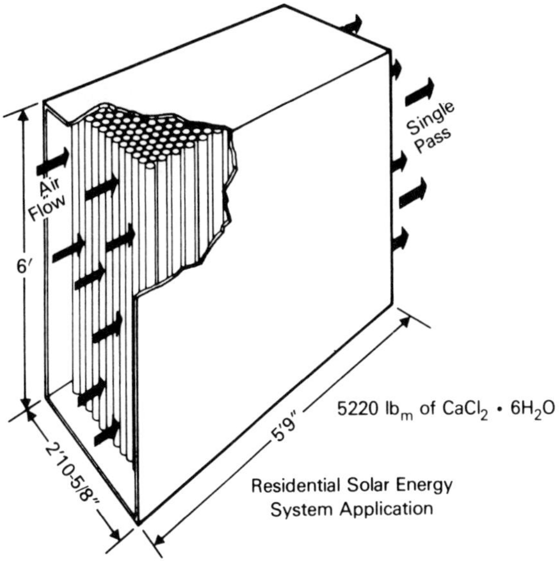

Fig. 6.5. "Energy storage rods" of PSI, Inc. with "Thermol 81" ($CaCl_2 . 6H_2O$, $T_f = 27.2°C$) in polyethylene (HDPE) cylinders [6.5]

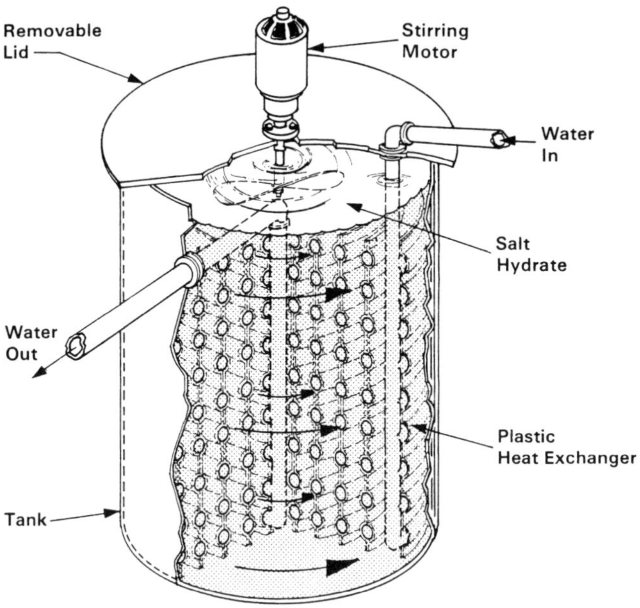

Fig. 6.6. CALMAC "heat bank" latent heat storage unit with $Na_2S_2O_3 . 5H_2O$, or $MgCl_2 . 6H_2O$ [6.5]

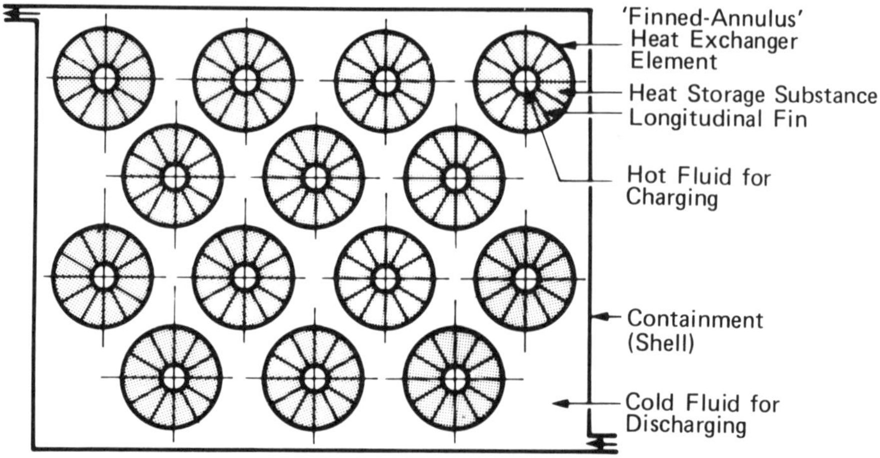

Fig. 6.7. The finned-annulus heat exchanger concept for latent heat storage [6.6]

6.4.5. Passive Systems

Solar heat gains are sometimes classified as "active" and "passive", the latter being a function of the building structure itself. The relevant properties of the building with respect to passive gains are mainly translucent exterior surfaces –windows or translucent covers–on the one hand and effective thermal capacity on the other. Suitably chosen TES properties will equalize internal temperature by damping the diurnal variations of insolation and ambient air temperature.

If the TES properties (thickness, thermal conductivity and heat capacity) of an exterior wall are properly chosen, then the diurnal oscillation of the effective outer surface temperature may be delayed by about 12 hours in passing through the wall, thus giving a welcome contribution to the heat balance of the room in the heating as well as in the cooling mode.

6.5. Heat Pumps

6.5.1. General

Performance control of most present-day heat pumps is achieved by on-off operation, much in the same way as in heating boilers (cf. Section 6.2). Standstill losses of heat pumps will be smaller than that of heating boilers since the supply temperatures are lower, only part of the system is at supply temperature, and there are no stack losses. However, it does take some time until the heat pump process has reached design conditions. During that time, the COP only slowly approaches its design value. Thus, start-up and shut-down losses may be even larger than for burners of heating boilers.

6.5.2. Buffer Storage

Time relays are often provided so as to reduce the frequency of switching. However, buffer storage on the supply side is advisable in almost all cases of residential heat pumps having water as heat transfer medium on the supply side [6.11]. For air-to-air heat pumps, a low frequency of switching can only be achieved by a relatively large temperature interval of the thermostat between the on and off temperatures, or by capacity control.

The buffer storage vessel is designed for displacement storage (Fig. 6.8) in order to minimize exergy losses by mixing that would affect the COP by increasing the heat pump outlet temperature at a given supply temperature. If the electric power supply of electrically-driven heat pumps is interrupted during the peak load hours of the day, buffer storage will bridge this gap in the same way as for electric boilers.

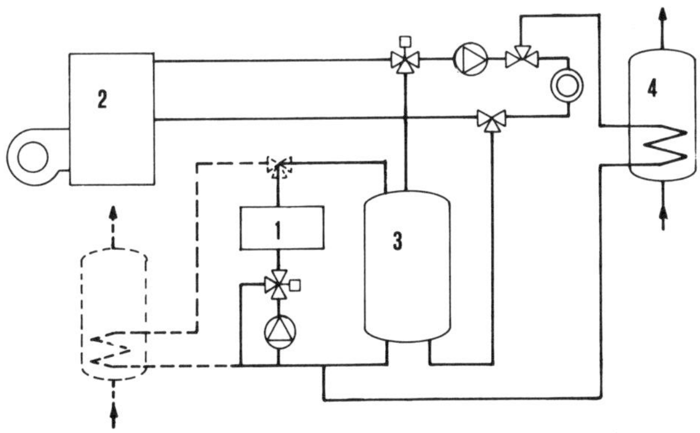

Fig. 6.8. Buffer storage for a heat pump [6.12]
1 air-to-water heat pump, *2* auxiliary boiler, *3* buffer store, *4* hot tap water store

Whereas speed control presently is rather expensive for electric heat pumps, it is less expensive for diesel or gas engine driven heat pumps [6.13]. On the other hand, speed control is possible only within limits and start-up and shut-down losses of engine-driven heat pumps are even higher than that of electric ones, because there are considerable losses of the engine in addition to the losses of the heat pump system. Therefore, buffer storage may even be useful for performance-controlled heat pumps.

Hot tap water production by means of heat pumps also requires buffer storage not only to reduce the frequency of switching but also to reduce the design power of the heat pump which otherwise would have to be designed for the high peak demand [6.13]. Heat transfer from the working medium to the store may be direct or indirect (see the left-hand and right-hand portions of Fig. 6.9).

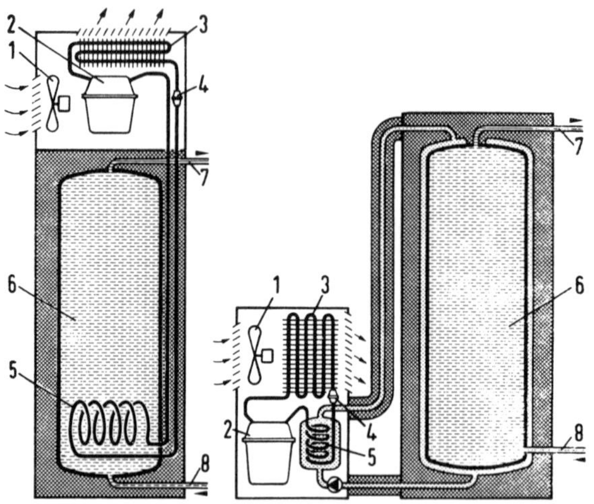

Fig. 6.9. Air-to-water heat pump for hot tap water production [6.14]
A package unit, *B* separate heat pump unit
1 blower, *2* compressor, *3* evaporator, *4* expansion valve, *5* condenser, *6* hot water store, *7* hot tap water supply, *8* cold water supply

6.5.3. Low-temperature Latent Heat Storage

For heat pumps having solar heat or ambient air as heat source, the need may arise for additional short-term or long-term storage on the low-temperature (heat-source) side [6.15].

Latent heat storage (ice storage) at 0°C is a very effective method of low-temperature storage. Ice storage systems have been developed and are now in use in several countries [6.12]. However, ice storage is particularly advantageous in climatic zones with cooling loads during the summer that are comparable to the heating loads in the winter (allowing annual cycling, that is, double use of the thermal storage system).

Fig. 6.10 shows the Annual Cycle Energy System (ACES) that was developed in the United States [6.16]. A heat pump supplies heat for a house from a water bin of 70 m^3. The water is frozen during winter and the accumulated cold serves for cooling purposes in summer. If all the ice is melted, the heat pump operates during the night, discharging heat through an outdoor coil and freezing ice for the next day's demand. The store acts as short-term cold storage in this mode of operation. In a northern climate, where heating requirements exceed those for cooling, the system requires a supplementary heat source. Simple

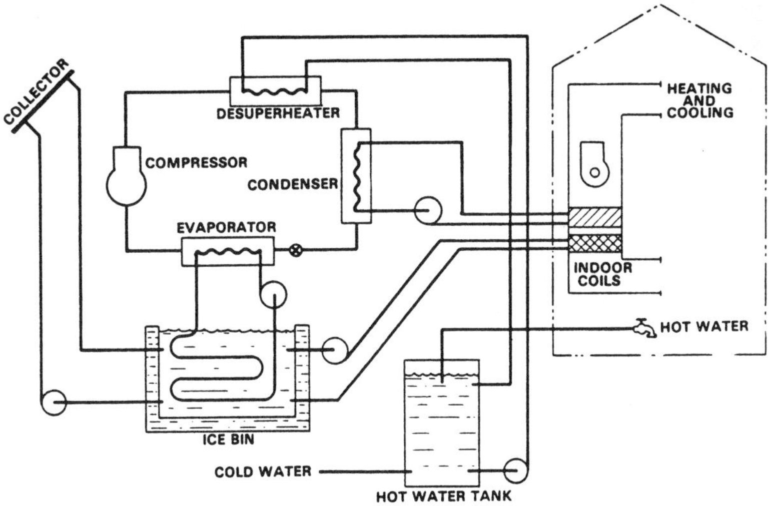

Fig. 6.10. Simplified diagram of annual cycle energy storage (ACES) [6.16]

unglazed solar panels can be used to satisfy this demand on account of the low temperature at which the panels have to operate. Domestic hot water is produced by desuperheating the vapor from the compressor. By such an arrangement water temperatures can be obtained which exceed the condensing temperature of the refrigerant. The experimental ACES project reports an annual saving of about 56% on electricity against a system based on full electric resistance heating in winter and electric air-conditioning in summer. A saving of 49% was found when a comparison was made with a system using an air-to-air heat pump instead of resistance heat in the control house of the project.

6.5.4. Low-temperature Subsoil Storage

Subsoil heat exchangers are another means of low-temperature heat storage. Shallow horizontal pipes in the ground provide a fair amount of storage but are essentially indirect heat exchangers to the ambient air.

Vertical subsoil heat exchangers (Fig. 6.11) provide increased storage for a much smaller surface area and with less earth movement (which is important in urban areas). However, here there is little direct heat exchange to the ambient and to the surrounding subsoil and there is, therefore, the need of recharging the store during the summer with solar energy or heat from ambient air, unless ground water is present for recharging [1.32].

6.5.5. Concentration Difference Energy Storage

The use of an intermittent, semi-closed, dry absorption heat pump with thermal storage by means of concentration difference rather than temperature

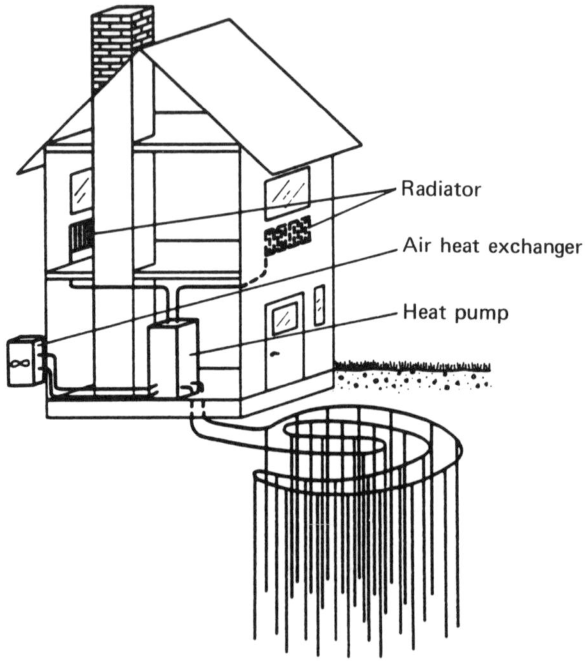

Fig. 6.11. Vertical earth heat exchanger for energy storage in the subsoil, heat being recharged from an air heat exchanger [6.17]

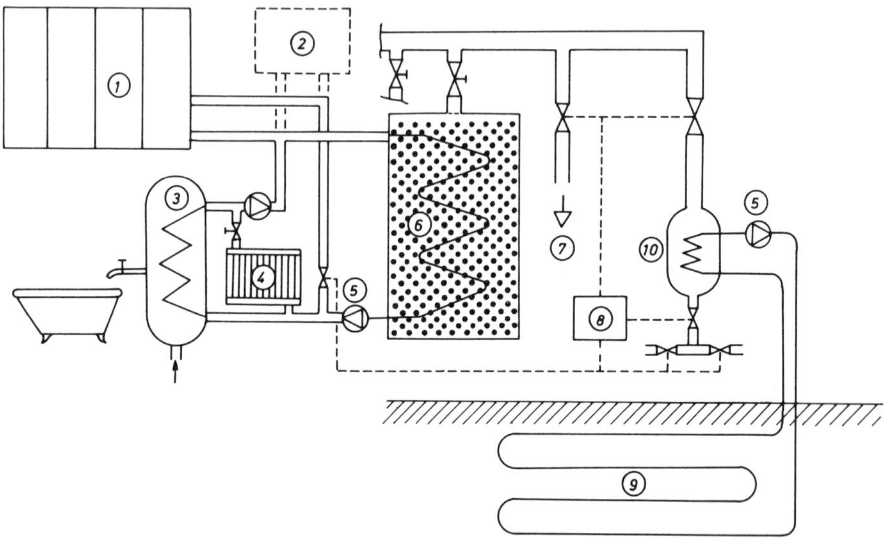

Fig. 6.12. Low temperature storage with absorption heat pump (Tepidus energy storing system, Sweden) in a single-family house [6.18]
1 solar collector, *2* alternative heat source, *3* tapwater, *4* radiator, *5* pump, *6* energy accumulator (Na_2S), *7* low pressure device, *8* control unit, *9* soil heat exchanger, *10* condenser/evaporator/water tank

difference (that is, at room temperature) and low temperature thermal storage in the ground is shown in Fig. 6.12. H_2O is the working fluid, Na_2S is the absorbent (7000 kg, storing approximately 7000 kWh), solar heat (75–100°C) from 40 m^2 collectors is the high-temperature heat source, the ground is the low-temperature heat source, the supply temperature is 50–60°C, the charge power is 20 kW and the discharge power 10 kW.

As an alternative to charging the system by solar energy, movable storage units of the type described have been suggested for transporting heat to and from an industrial waste heat source [6.18].

Absorption heat pumps are often called "Chemical Heat Pumps" which sometimes is abbreviated as CHP. However, this should not be confused with the abbreviation for combined heat and power (cogeneration) plants.

6.6. District Heating

6.6.1. General

The basic requirements of TES for district heating plants are similar to the requirements for individual heating plants. Fig. 6.13 shows a plot of annual load duration for district heating. The annual utilization factor of district heating, including the use for domestic hot water throughout the year, is rather low (34%; that is, about 3000 h/a). The plot shows that peak heating power is used only for a very short time and that about 90% of annual energy may be covered with an installation which rates 45% of maximum heating power [6.19].

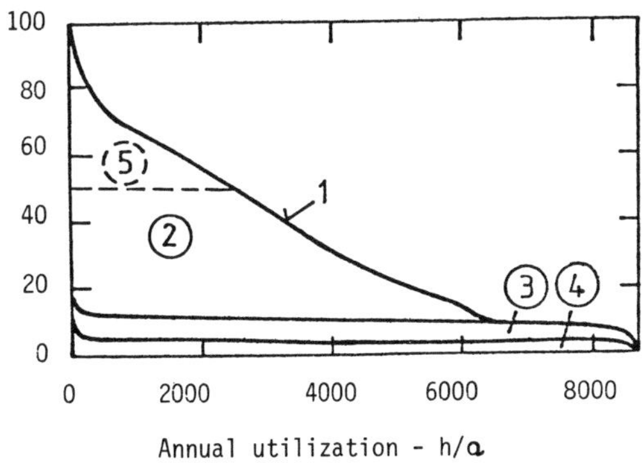

Fig. 6.13. Typical annual load duration for district heating [6.19]
1 total load, *2* space heating, *3* hot domestic water, *4* grid losses, *5* peak load

As has been shown in Chapter 1 (see Fig. 1.1 A), the TES plant may be located at the energy conversion plant or at the end of the long-distance piping. Storage at the consumer is expensive. The optimum location also depends on the type of the energy conversion plant and the type of the store as well as on the kind of operation of the grid and of the plant. High-efficiency energy conversion plants such as CHP plants and the long-distance piping are usually not designed for maximum heating power but for only 50 to 60% of it, the remainder being covered by peaking and/or storage plants at the end of the long-distance piping, before the heat distribution system. Depending on its capacity, such a storage plant will reduce the utilization of the peaking plant with its expensive fuel. Indeed, if it is a seasonal storage plant, it will remove the need for a peaking plant altogether. In the following, only short-term (daily or perhaps weekly) storage is considered, seasonal storage being dealt with in Section 6.7.

Hot water district heating grids may be operated as follows:

(a) with constant supply temperature, the flow rate depending on heat demand;
(b) with sliding supply temperature (depending on heat demand or on ambient temperature) and constant flow rate (at least down to the minimum temperature of 60 to 70°C necessary for domestic hot water);
(c) by a combination of (a) and (b) above, that is, with a constant-temperature supply line, a sliding-temperature supply line and a common return line.

Hot water being the heat transport medium in heat supply systems, it is used as storage medium too. Contrary to industrial applications, displacement storage is used rather than sliding-pressure storage. Corresponding to supply temperatures of between 90 to 180°C, design pressures including static heads are below 30 bar. Unpressurized vessels for temperatures of up to 95°C (Fig. 6.14) are, of course, less expensive [6.20].

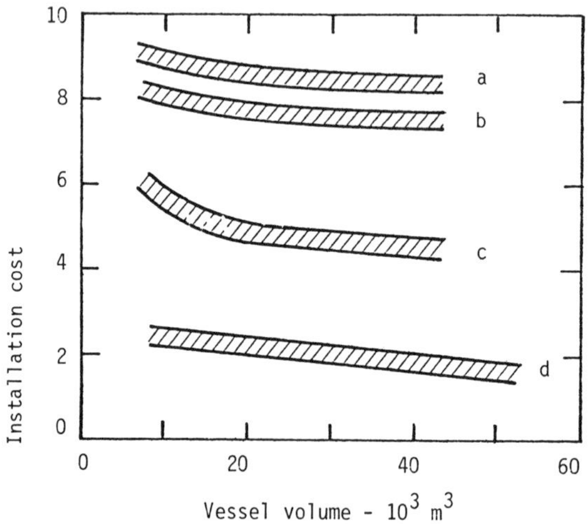

Fig. 6.14. Specific installation costs of pressurized and unpressurized hot water accumulators (in DM/kWh) as a function of vessel volume [4.27]
a cylindrical pressure vessel 21 bar (200/50°C), *b* cylindrical pressure vessel 11 bar (170/50°C), *c* spherical pressure vessel 11 bar (170/50°C), *d* unpressurized tanks (90/50°C)

In order to increase storage capacity, operation should always be as near to design temperature as possible. If storage vessels are arranged in sliding-temperature grids, operation at design temperature may be advantageous (in spite of the additional exergy losses) even if the instantaneous supply temperature is lower.

6.6.2. Types and Operation of Heating Plant and Heat Storage

The heat load of district heating plants may be covered by pure heating plants, by waste heat utilization plants or by combined heat and power generation (CHP; cogeneration) plants.

The main advantage of CHP is high fuel efficiency. If heat supply is the primary consideration, electric power may be generated in a topping plant having very high efficiency. If, on the other hand, power generation is the primary concern, then the "waste heat" of the power may be used for heating purposes.

It should be noted, however, that only with gas turbine plants the waste heat is of sufficiently high temperature to be useful for space heating purposes. In steam power plants, it is necessary to increase the temperature of the "waste heat" to be useful from the usual cooling water outlet temperature of about 15 to 30°C to the actual district heating supply temperature. This, of course, entails either a reduction of electric power generation (usually at about 10 to 20% of heat production) at given fuel input and boiler output or, at constant turbine power and sufficient boiler capacity, an increase of fuel consumption and steam production (this usually amounts to about 30 to 60% of heat production, depending mainly on the district heating supply and the return temperatures, load and number of heating condenser stages).

Fig. 6.15 I shows a daily heat and power load diagram of a typical town. It is obvious that the smoothing of heat peaks and decoupling of heat and power in a CHP plant by means of thermal energy storage will enhance the flexibility of the system and improve its use.

For daily storage, the following cases [6.20, 6.22] are to be considered:

(a) Equalization of daily heat load so as to achieve constant load at the heating or CHP plant (Fig. 6.15 A): For better utilization not only of the energy conversion plant but of also the long-distance piping, the storage plant should be at the end of the long-distance piping (see Fig. 1.1, Case d). A heat analysis for the district heating systems in seven German towns [6.20] has shown that maximum storage capacity for daily storage is 0.8 to 1.5 maximum heat load hours and is required at ambient temperatures of about 0°C (below that temperature load variations are less because of increased night heating). The maximum charge and discharge power is 0.1 to 0.3 times maximum heat load and occurs at ambient temperatures of from − 5 to + 5°C.

(b) Modulation of the heat load profile according to the power demand profile for CHP plants of the topping type (that is, without low temperature condenser), covering electrical peak and medium load (Fig. 6.15 B): If the charging load is larger than the heat load, the storage plant should be situated at the CHP plant (see Fig. 1.1, Case a).

(c) Modulation of electrical power from a bleeding steam CHP plant for maximum electricity production at peak demand periods: The storage plant is located at the CHP plant. It is charged during periods of low electric demand (for instance, between 9 p.m. and 6 a.m.) and discharged during the peak load periods (say, between 6 a.m. and 9 p.m.), thus ensuring constant heat supply (Fig. 6.15 C). The required storage capacity is 8.2 to 13.2 peak heat load hours and the maximum charge power is 0.85 to 1.0 of peak heat load. However, both are much higher than for Case (a). The storage temperature is either equal to the supply temperature or is around 95°C, with unpressurized vessels and boosting by the CHP plant as required.

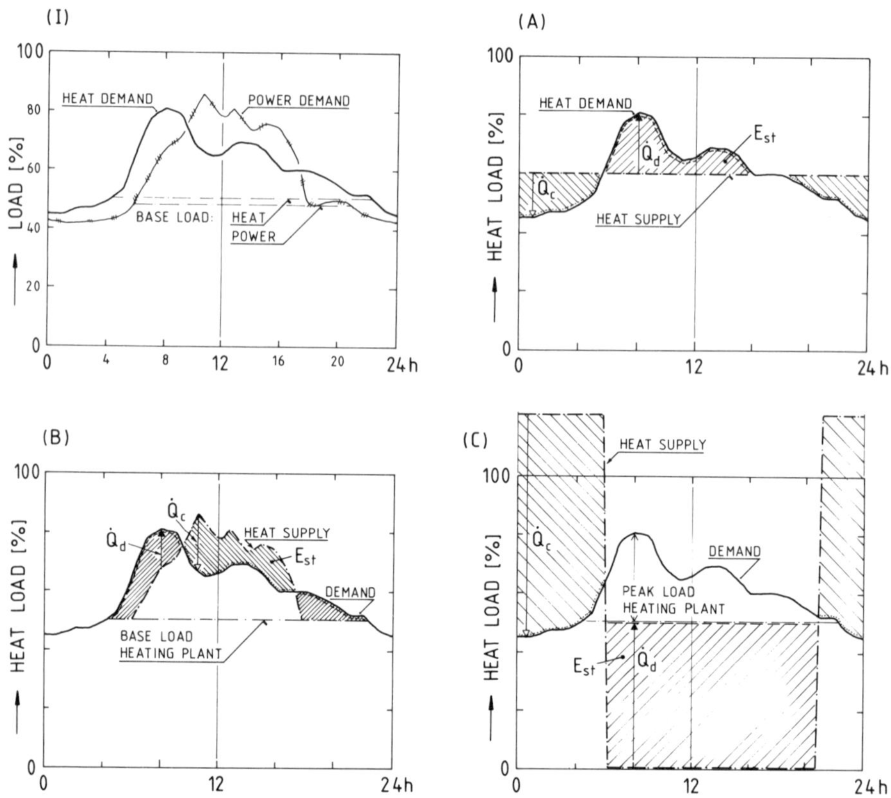

Fig. 6.15. Modes of storage operation [6.20]
I daily heat load and power load diagram, *A* equalization of heat load, *B* peak power generation with topping CHP plants, *C* trapezoidal power generation with bleed CHP plants, $\dot{Q}_c$ charge load, $\dot{Q}_d$ discharge load, E_{st} stored heat

(d) Using long-distance piping as store means some exergy loss but requires almost no additional investment. The supply as well as the return line may be used for thermal storage. Supply line storage consists of increasing the supply line temperature and keeping the water flow rather constant. There is a mixing station (Fig. 6.16) at the end of the long-distance line. When the water with increased temperature arrives (say, after about 2 hours, if the water velocity is 3 m/s and the length is 22 km), the return line water is admixed to keep the supply temperature to the consumers constant. At the same time, the water flow in the long-distance pipe is reduced so as to keep the water flow in the supply line to the consumers constant. The store is now charged. It may be discharged by reducing the supply line temperature to normal design value at reduced water flow. The return line may be charged and discharged in a similar manner, using the charging pump (see Fig. 6.16). Both methods require some flexibility of the CHP plant and some measuring and control equipment so that the operator may always be aware of the actual state of the miles of the store and can act accordingly.

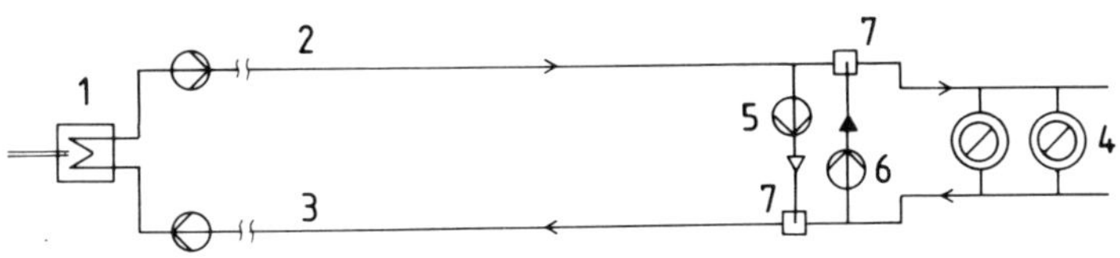

Fig. 6.16. Storage in long-distance piping
1 heat source, *2* supply line, *3* return line, *4* consumers, *5* charge pump, *6* discharge pump, *7* mixing stations

Table 6.2. *Data of a Waste Incineration Plant of the Vienna Municipal District Heating Company*

(a) Power Generation	
Auxiliary Steam Turbine	2 MW
(b) Heat Generation	
2 Incineration Boilers	2 × (20–36 MW)
2 Oil-fired Boilers (Peak)	2 × 52 MW
3 Oil-fired Boilers (Reserve)	3 × 19 MW
(c) Maximum Net Output of Heat	230 MW
(d) Water Temperature (Supply)	150°C
(Return)	75°C
(e) Storage Farm	
Type	vertical, displacement
Number of Vessels	9
Diameter	3.1 m
Height	22 m
Pressure (Operation/Design)	20/25 bar
Volume per Vessel	157 m^3
Total Volume	1413 m^3
Storage Capacity, Heat	120 MWh

6.6.3. Hot Water Storage in a Waste Incineration Plant

Table 6.2 shows data from a heating plant of the Vienna municipal district heating company. There are two incinerating boilers for the heating base load, two oil-fired boilers for the additional peaks and three small boilers as reserve units.

The discrepancies between the fluctuating supply of heat from the incinerating boilers and the daily fluctuation of heat demand is matched by 9 displacement accumulators of 157 m^3 each. Since incineration continues through the year, the surplus heat during summer has to be rejected, this rejection being accomplished by means of a heat exchanger that is cooled by water from the river.

6.6.4. Hot Water Storage in Gas Turbine CHP Plants

Open-cycle gas turbine plants show low investment cost and are able to start up relatively quickly (10 to 15 minutes usually and about 6 minutes for quick start-up). However, their efficiency is only moderate (about 27%). The fuel used is light fuel oil or natural gas and therefore fuel cost is high. The plants are restricted to peak load duty (500 to 2000 h/a). Waste gas temperature is of the order of 500°C, permitting the addition of a heating boiler for district heating without reduction of electric power (except for overcoming the relatively small gas pressure drop in the boiler). Normal operation of such a CHP plant is based on electric power demand according to Case (b) of Section 6.6.2 and Fig. 6.15 B. The ratio of electrical power to heat is high and fuel utilization depends strongly on the load (see Fig. 6.17).

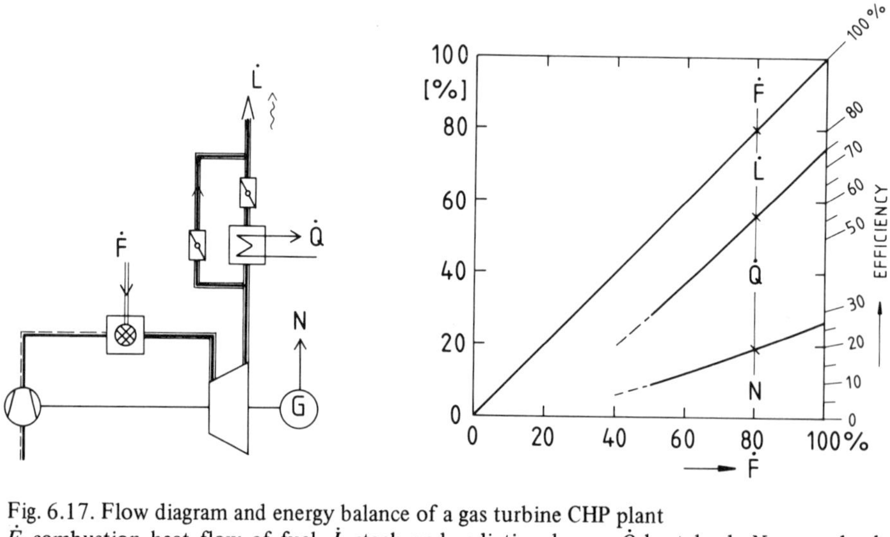

Fig. 6.17. Flow diagram and energy balance of a gas turbine CHP plant
$\dot{F}$ combustion heat flow of fuel, $\dot{L}$ stack and radiation losses, $\dot{Q}$ heat load, N power load, $N/\dot{F}$ thermal efficiency, $(\dot{Q}+N)/\dot{F}$ fuel utilization

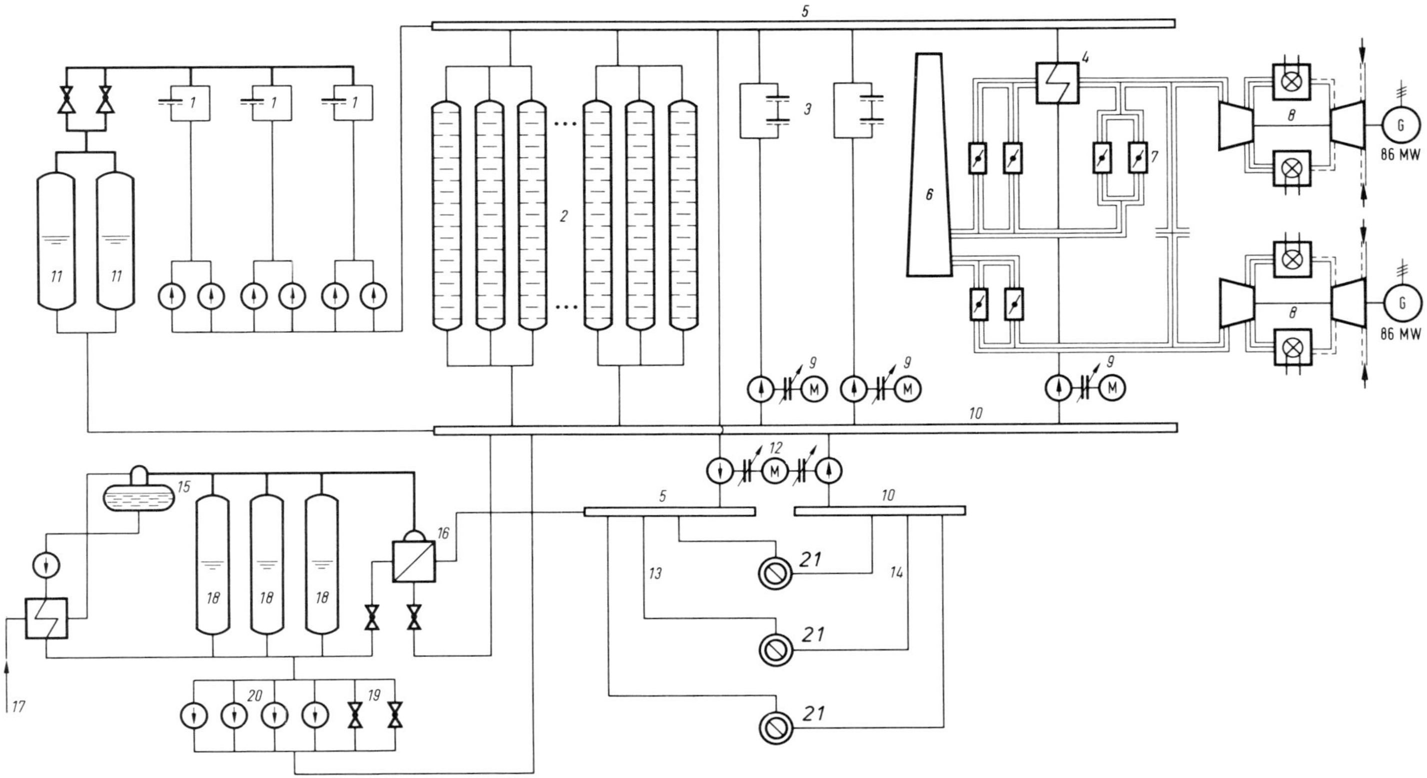

Fig. 6.18. Flow sheet of the CHP plant Munich-Freimann with gas turbines and displacement storage [6.21]
1 pressurizer boilers, *2* hot water storage vessels (15 × 310 = 4650 m^3), *3* hot water boilers, *4* waste heat boiler, *5* supply header, *6* chimney, *7* bypass, *8* gas turbines, *9* feed pumps, *10* return header, *11* expansion vessel, *12* recirculation pumps (four units), *13* supply line, *14* return line, *15* make-up water tank + degasser, *16* steam generator 1.5 bar, *17* make-up water supply, *18* compensating vessels, *19* return valves, *20* grid make-up pumps, *21* consumers

Fig. 6.18 shows the flow sheet of the CHP plant at Freimann (near Munich in the Federal Republic of Germany) which has been operable since 1972. There are two gas turbine sets of 80 and 86 MW electric power. One operates as a CHP plant, while the other generates electric peak load only. Two heating boilers and 15 vertical cylindrical storage vessels are connected in parallel to the gas turbine waste heat boiler unit. Further design data of this plant are given in Table 6.3.

Table 6.3. *Data of the CHP Plant Freimann near Munich* [6.21]

(a)	Power Generation	
	Gas Turbine No. 1	80 MW
	Gas Turbine No. 2 (Peak only)	86 MW
	Total Power Generation	166 MW
(b)	Heat Generation	
	Waste Heat Boiler behind Gas Turbine No. 1	136–145 MW
	Gas/Oil-fired Boiler No. 1 (Peak)	134 MW
	Gas/Oil-fired Boiler No. 2 (Reserve)	(134 MW)
(c)	Maximum Net Output of Heat	307 MW
(d)	Water Temperature (Supply)	140–180°C
	(Return)	60–80°C
(e)	Storage Farm	
	Type	vertical, displacement
	Number of Vessels	15
	Diameter	3.6 m
	Height	33 m
	Design Pressure	27 bar
	Volume per Vessel	310 m^3
	Total Volume	4650 m^3
	Storage Capacity (160/70°C), Heat	500 MWh

6.6.5. Hot Water Storage in Steam CHP Topping Plants

The steam CHP topping plant has already been mentioned in Section 5.3.3. Steam expansion in the turbine is limited to a saturation temperature given by the supply line temperature plus heating condenser temperature difference. If water heating takes place in two or more condenser stages, the first one (return temperature) is heated by back-pressure steam, while the following ones are heated by bleed steam. Alternatively, the topping turbine may have two casings with different back-pressures. The steam from the longer one heats the first, the steam from the shorter one heats the second stage of the water heater. Steam parameters of such plants are not extreme (that is, below 540°C, 140 bar and without reheat).

Contrary to the open-cycle gas turbine CHP plant, the steam CHP topping plant cannot be operated without heat load since it has no heat sink except the district heating system. Thermal storage–Case (a) or (b) of Section 6.6.2–will therefore be essential for decoupling electricity and heat production.

The ratio of electric to heat power is lower than that of a gas turbine CHP (it also depends on the supply temperature and–with two or more heater units –even on return temperature), but fuel utilization and part load behaviour is better (Fig. 6.19) [6.21]. Peak load boilers should be connected in series after the CHP plant.

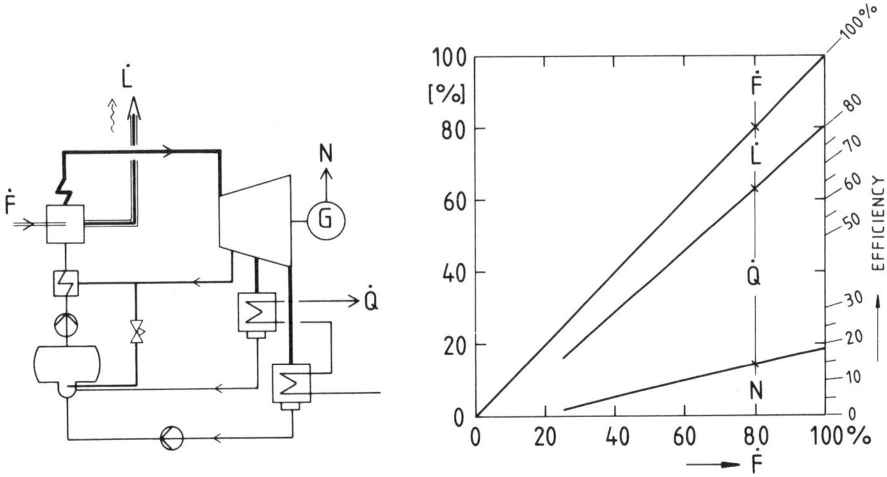

Fig. 6.19. Flow diagram and energy balance of a steam CHP plant (for symbols see Fig. 6.17)

6.6.6. Hot Water Storage in Bleeding Steam CHP Plants

Fig. 6.20 shows the flow sheet of a bleeding steam (steam extraction) CHP plant. There is only one stage of bleed steam heaters. Operation is with sliding temperature; peak supply temperature is achieved by throttling before the LP stage [3.6]. Fig. 6.21 shows the simplified characteristic, that is, the electric power and heat load as a function of thermal power of fuel.

Operation of such plants is in accordance with Cases (a) or (c) of Section 6.6.2 and Figs. 6.15 A or C. In the latter case, the plant becomes a peak load plant. For short peaks, the storage capacity required is small and the storage plant will be economical. The store is charged during low loads on the electric and the heating side and is discharged during periods of peak electric demand, thus permitting the utilization of the full electric power. Without storage, the loss of electric power at maximum heat load would have been 17%.

Economic conditions in steam CHP plants with bleeding may be quite different. Thermal storage will be economic in many but not in all cases [6.21]. For instance, if the boiler is designed so as to be able to supply steam for the full electrical power plus full (or a large portion of) heating load, the need for thermal storage will not arise (or will be greatly reduced).

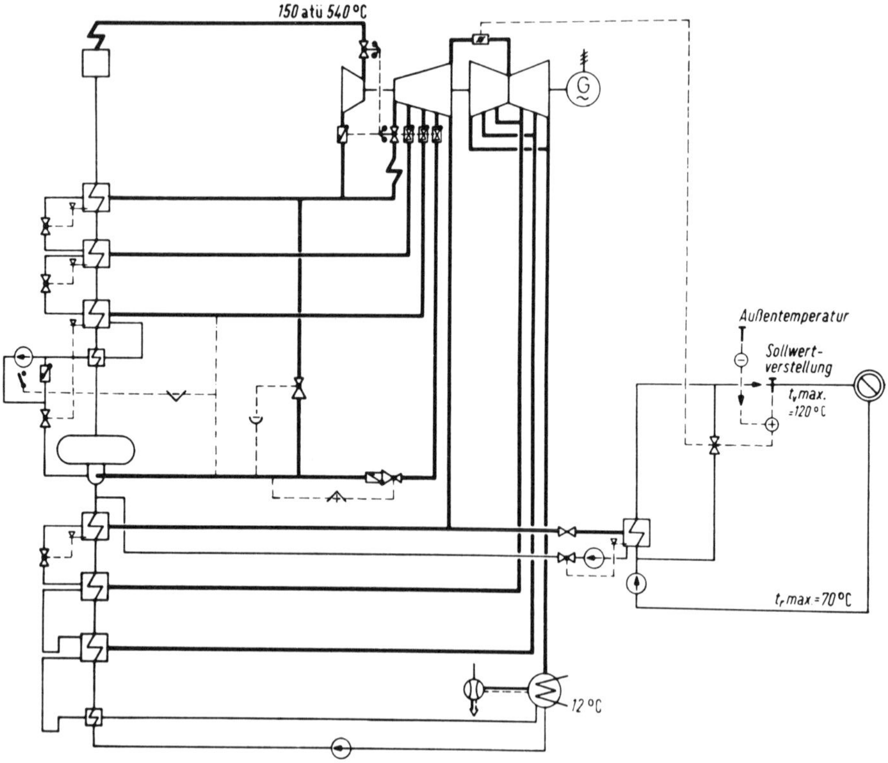

Fig. 6.20. Flow diagram of a bleeding steam CHP plant [3.6]

6.6.7. Hot Water Storage in Combined Gas Turbine and Steam CHP Plants

In recent years, several CHP plants based on the combined gas turbine and steam process have been built in the Federal Republic of Germany (South Munich, Bielefeld, Saarbrücken). The advantage of this type of plant is the high ratio of electric to thermal power. It requires, of course, a fuel which is suitable for the gas turbine.

The flow sheet of the steam part is usually of the bleeding steam (steam extraction) type described in the previous section. Accordingly, the arrangement and operation of hot water storage (if any) is similar to this type of plant. For instance, the Bielefeld CHP plant (which has a 25 MW gas turbine and a total of 85 MW steam turbine power) has a hot water storage capacity of $8 \times 400 = 3200$ m^3 [6.23].

6.6.8. Combined District Heating and Feed Water Storage

The discussions of TES in bleeding steam CHP plants given in Section 6.6.6 and of feed water storage given in Section 7.4.2 show several parallels. Both

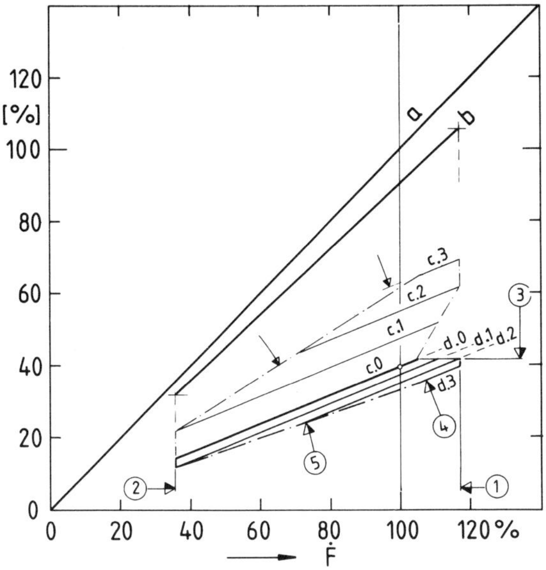

Fig. 6.21. Characteristic of a bleeding steam CHP plant (flow of energy versus combustion heat flow of fuel)
a combustion heat flow of fuel ($\dot{F}$), ($a - b$) stack and radiation losses ($\dot{L}$), *b* boiler net output ($\dot{C} + \dot{Q} + N$), *c.0* = *d.0* power (N) without heat load (pure condensing operation), *c.1* to *c.3* sum of heat and power ($\dot{Q} + N$) for heat loads from 0.1 to 0.3 ($\dot{F}$), *d.1* to *d.3* power load for heat loads from 0.1 to 0.3 ($\dot{F}$), ($b - c$) load $\dot{C}$ of condenser, ($c - d$) heat load ($\dot{Q}$). Limitations of operation: 1. maximum fuel flow or boiler load, 2. minimum fuel flow or boiler load, 3. maximum load of turboset, 4. maximum load of heating condenser, 5. maximum bleed ratio of CHP plant

methods allow peak load coverage and in both cases hot water, heated by bleed steam, is stored. The only difference is that the water is used as heating water in one case and as feed water in the other. A combination of the two processes appears possible and promising.

6.7. Large-scale Seasonal Thermal Storage

Solar heating at high latitudes requires seasonal thermal storage. Moreover, as was pointed out in the previous chapters, seasonal storage may also be of interest for heat pump, waste heat and CHP applications.

Large-scale vessels for low temperature seasonal storage have been discussed in Chapter 4. The steel membrane reservoir (Fig. 4.26), the rock cavern (Fig. 4.27) and lake storage (Fig. 4.29) are significant in this respect.

A method under development for large-scale seasonal storage is the multiple well system [6.24]. It consists of boreholes in rock or clay with water circulation for charging and discharging (Fig. 6.22). This system is similar to the subsoil storage system with vertical subsoil heat exchangers for heat pump operation as

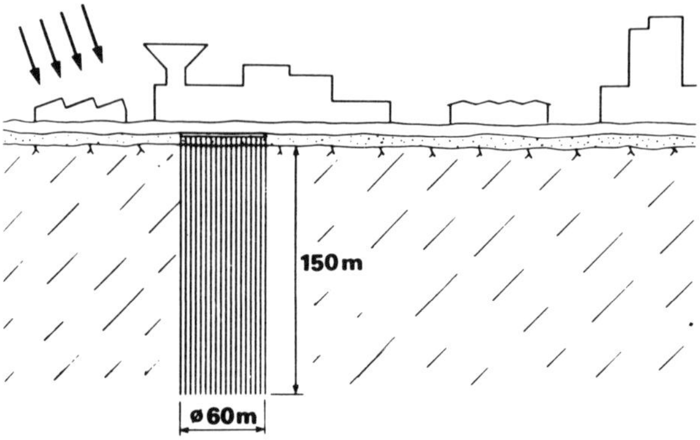

Fig. 6.22. Multiple well system for heat storage in hard rock [6.24]

mentioned in Section 6.5.3. However, the surface losses to ambient and to neighbouring subsoil are relatively less important at the very large sizes of the store and higher storage temperatures are therefore feasible.

A further method is aquifer storage where an aquifer between two layers of confining beds is charged and discharged via a well injection and withdrawal system (Fig. 6.23).

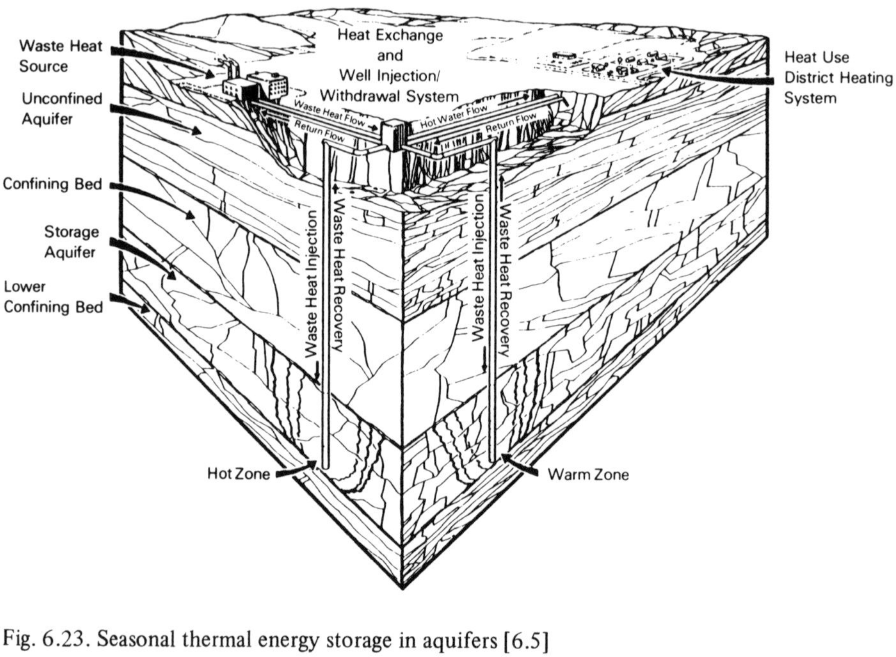

Fig. 6.23. Seasonal thermal energy storage in aquifers [6.5]

6.8. Conclusions

There are four main incentives for TES that are of increasing importance in space heating and cooling:

- improved seasonal efficiency of boilers and heat pumps;
- peak lopping of electricity demand by means of off-peak storage heaters;
- increased demand coverage by solar installations;
- decoupling of heat and power generation in CHP plants.

TES is mostly short-term (a few hours to a few days). Unpressurized or pressurized water is used wherever there is a water system for heat distribution. Large-scale as well as small-scale TES equipment is in wide-spread use.

Whether long-term—and in particular seasonal—TES for heating and cooling applications is or will become cost-effective, is an open question. The economic barrier is a stringent one since the seasonal storage by definition is used only once a year and its first cost therefore must typically be less than about 10 times the value of its energy content in order to be cost-effective. Such a criterium will be met by a large-scale seasonal store more easily than by a small-scale one. Also specific losses will generally be smaller for the large-scale store. The emphasis of R, D & D is therefore on the large stores. Until recently maximum store sizes were a few hundred m^3 per vessel and few thousand m^3 per plant. However, these figures are 10^4 to 10^5 m^3 per storage unit today.

7. Power Plants with Thermal Energy Storage

7.1. General

As was pointed out in detail in Chapter 1, energy storage for electric power generation (flow energy storage or pumped storage) may serve one or more of the following purposes:

- economic peak load coverage;
- bottoming;
- stop-gap duty, either by quick start-up or by idling of turbomachinery;
- short-term storage for facilitating control by buffering (that is, integrated storage for increasing spinning reserve).

The advantages of energy storage include:

- lower initial cost (replacing separate peak load plants by better utilization of base load plants);
- lower fuel cost (of base load plants versus peak load plants) and oil substitution;
- potentially better overall efficiency than that of peak load plants;
- operational improvements (less cycling of fuel elements in a nuclear reactor; avoidance of extreme part loads);
- environmental improvements (if the base load plant is of a low emission type or if it is better equipped with emission control).

In the following, pumped thermal storage from the grid (that is, steam pumped storage, pneumatic storage, compressed air energy storage with additional firing and gas turbine quick start-up storage) will be dealt with first. This is then followed by a look at integrated pumped storage (shaft-driven) and integrated flow storage according to fuel (coal, nuclear, solar) and process (steam and gas turbine).

7.2. Steam Pumped Storage

The concept of steam pumped storage (as was explained in Section 3.1.6) was proposed by Marguerre [3.4]. A peak load power plant of 15 MW with 2 hours full load discharge duration (4 hours for triangular load) and a steam compressor power of 7.5 MW (that is, a charging ratio of 1:2) was designed and apparently built.

A peak load plant of this kind does not discharge any effluents to the environment, since it burns no fuel. The low pressure storage vessel acts as heat sink during peak load operation. Using above-ground vessels, it may therefore be located at any site either at or near the centers of demand. In this way, not only the base load plant is better utilized but the grid also.

Fig. 7.1 gives the flow sheet of the plant mentioned in the above discussion. Charging is achieved by flashing hot water in several stages from the low-pressure storage vessel. The low-pressure steam is then compressed and superheated in the multi-stage steam compressor. The high-pressure vessels are designed for only 13 bar according to the technology available during the 1930s. There is no intermediate cooling although the superheat in the high-pressure stages is reduced by

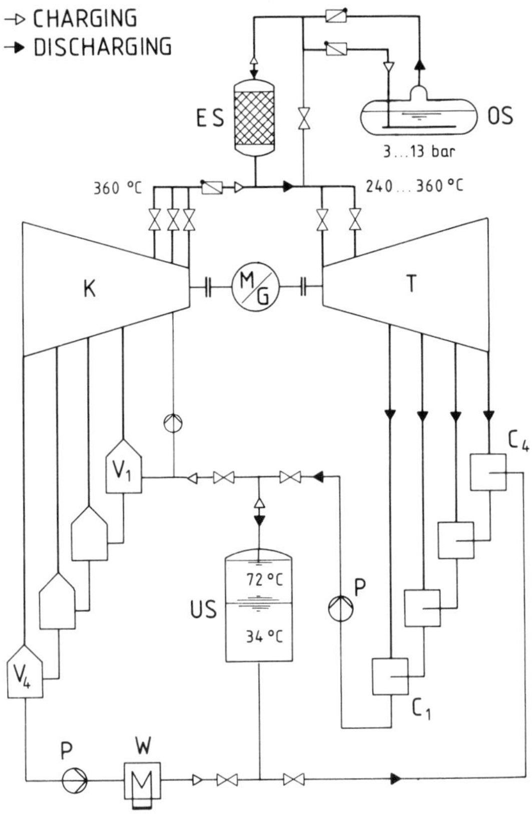

Fig. 7.1. Flow sheet of a steam pumped storage plant [3.4]
MG motor/generator, *K* compressor, *T* steam turbine, *OS* upper pressure storage, *ES* superheat storage, *US* lower pressure storage, V_1 to V_4 flash evaporator, C_1 to C_4 condenser, *W* heat sink, *P* pumps

water sprays. However, the superheat energy of the last stage is stored in a superheat storage vessel with oil or solid bodies as storage media. For discharging, the motor/generator set is coupled to the turbine. Exhaust steam and bleed steam are condensed in mixing condensers, thus heating up the cold water from the low-pressure storage vessel.

The storage efficiency (turnaround efficiency) of this system is rather low (about 50%). This is so mainly because of the exergy losses in compressor and turbine. In this closed system, an excess of low temperature heat is therefore generated, an excess which must be discharged to ambient. Alternatively, it may be used for district heating schemes.

7.3. Storage in Gas Turbine Power Plants

7.3.1. Pneumatic Pumped Storage

The pumped-storage principle may be applied to air as the working and storage medium (see Section 3.1.5). This means a simplification, since the

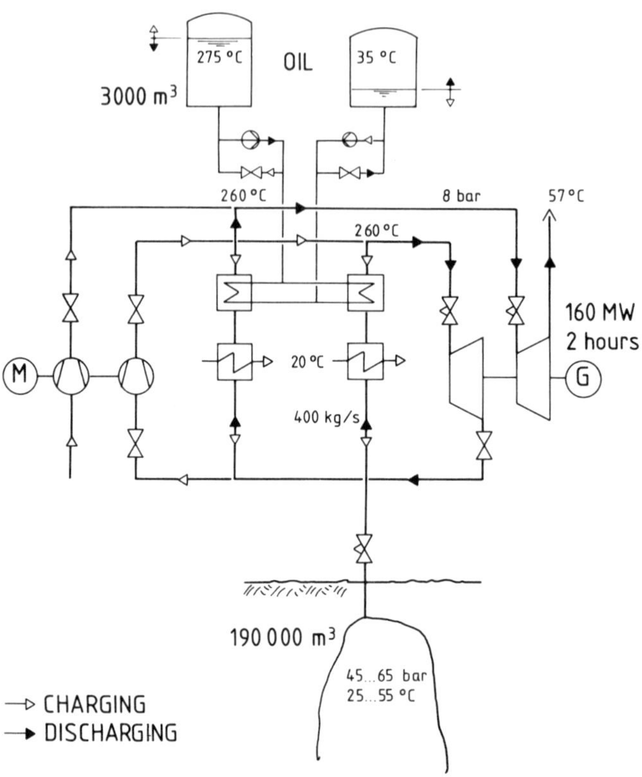

Fig. 7.2. Pneumatic pumped storage [7.1]

atmosphere may now function as low pressure storage. Fig. 7.2 shows the flow sheet of such a proposed plant [7.1] having a 160 MW peak load power and 2 hours full load discharge capability. There is a two-stage compressor for charging the air storage vessel. The air temperature rises in the LP stage but is reduced in an intermediate air cooler so as to minimize the power input to the HP stage. After the HP stage (that is, before entering the storage vessel), the air is cooled again, this time in order to increase the capacity of the storage vessel. However, the heat taken up in the two air coolers is not (or perhaps only to some extent) discharged to ambient. Rather it is stored in a TES system. In the case considered, this is a two-vessel system with oil as storage medium. In Fig. 7.2, the thermal conditions and flow rates for discharge operation are shown.

Turnaround efficiency is about 65 to 70%. It is higher than that of the steam pumped storage system, because the pressure variations are small in the storage system (upper pressure) and are zero in case of the atmosphere acting as "lower pressure system". Nevertheless, the storage density is rather low and an inexpensive underground storage vessel is therefore required. In the case shown, it is a salt cavern acting as a sliding-pressure system. Thus, the system is–contrary to the steam pumped-storage system–restricted to special locations.

The pneumatic pumped storage system may be used not only for programmed peak load shaving but also as (almost) instantaneous reserve plant, since start-up time may be rather short as compared to a thermal power plant.

The pneumatic pumped storage plant has higher initial cost than a compressed air energy storage (CAES) plant. It also requires more (electric off-peak) pumping power, but no fuel is needed. Thus, it will be economic only if fuel cost is very high and off-peak electricity rates are low [7.1, 7.2].

The principle of pneumatic pumped storage (with separate thermal energy storage of compressor heat) may be combined with the open-cycle gas turbine based on fossil fuel or on solar energy. In these situations, not only the compressor heat but also high temperature heat from the cycle may be stored. For a solar gas (air) turbine power plant, an underground thermal store with granite pebbles of about 1/2 in. diameter, working temperatures between about 500 and 800°C, in addition to a pressurized air store, has been suggested [7.3].

7.3.2. Compressed Air Energy Storage (CAES)

The Compressed Air Energy Storage (CAES) system (see Section 3.1.4) may be considered as a pneumatic pumped storage system with a furnace chamber (which is fuelled with oil or gas) between storage vessel and turbine. The working fluid is now an air/flue-gas mixture. Turbine power will increase approximately proportionally to the absolute gas temperature at the turbine inlet.

Alternatively, the CAES system may be considered as an open-cycle gas turbine plant with an air storage system after the compressor, decoupling the working times of compressor (charging period) and turbine (discharging period).

Such a plant has the advantage of being able to run indefinitely as a power plant and of being suited for discharge operation with about 3 times the power of normal operation. However, optimization studies [7.4] show that the optimum storage pressure is between 40 and 90 bar and this is far above the optimum pressure for open-cycle gas turbines. Operation without the storage system would therefore entail low efficiency. Furthermore, initial cost may be considerably reduced if the charging power of the compressor is kept low. Today a charge/discharge power ratio of 1:4 is considered optimal (that is, charge power is 25% of discharge power). In this way, the CAES plant looses the ability of running without the storage system; it becomes a peaking plant, suited for programmed peaks of a few hours per day, or a reserve plant with a start-up time of a few minutes; it is not suited for instantaneous reserve, however.

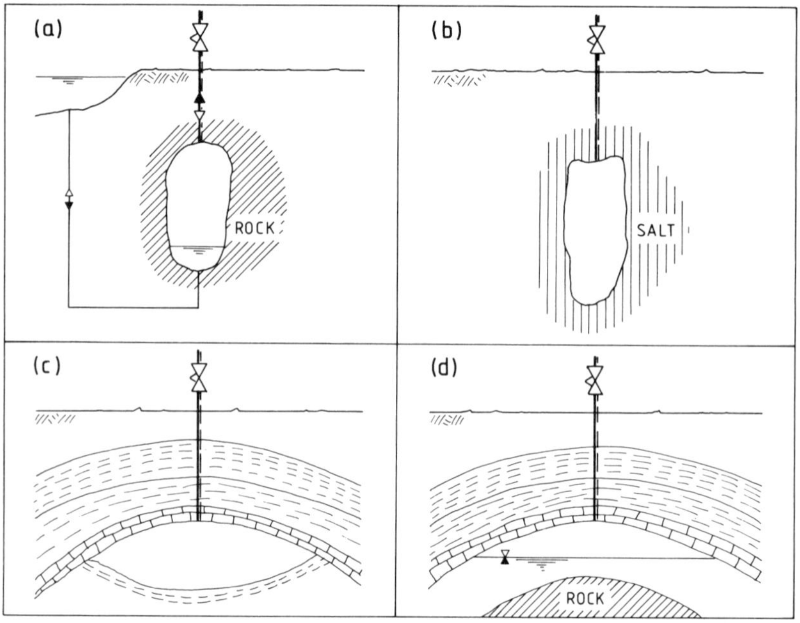

Fig. 7.3. Four subterranean methods for compressed air storage [7.33]
a mined cavern with hydrostatic compensation, *b* solution-mined salt cavity, *c* depleted oil or natural gas field, *d* porous water-filled structure (aquifer)

Even if CAES plants require somewhat less storage volume than pneumatic pumped storage plants, their storage density is still low (about 2 kWh/m^3), thus they too require inexpensive underground storage vessels. Four possibilities (Fig. 7.3) have been proposed in this respect [7.5]:

(a) a rock cavern with a water basin acting as constant pressure (displacement) storage;
(b) an artificial salt dome acting as a sliding-pressure system;
(c) a vacant (used) natural gas or oil field acting as a sliding-pressure system;
(d) an aquifer with an impermeable lid acting as displacement storage.

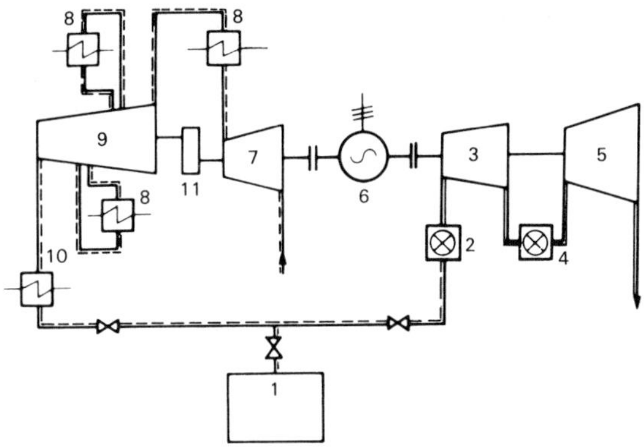

Fig. 7.4. Flow sheet of the compressed air energy storage (CAES) plant Huntorf [7.6]
1 storage cavern, *2* HP furnace, *3* HP turbine, *4* LP furnace, *5* LP turbine, *6* generator/motor, *7* compressor, *8* intermediate cooler, *9* HP compressor, *10* after-cooler, *11* gearbox

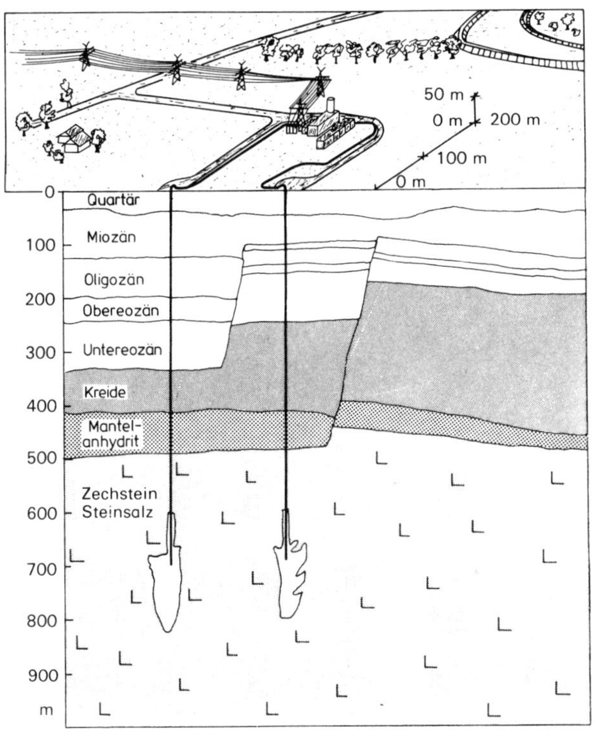

Fig. 7.5. CAES plant Huntorf–the geological situation [7.6]

Built at Huntorf near Bremen in the Federal Republic of Germany for Nordwestdeutsche Kraftwerke AG, Hamburg (NWK), the first CAES plant has been in operation since 1978 [7.6]. Fig. 7.4 gives the flow sheet. The motor/

Table 7.1. *Main Data of Compressed Air Energy Storage (CAES) Plant Huntorf* [7.6]

1. *Gas Turbine*	
Type	single-shaft, with reheat
Power	290 MW
Speed	3000/minute
Air Flow	417 kg/s
Flue Gas at Turbine Inlet	
HP	approx. 40 bar/550°C
LP	approx. 10 bar/825°C
Waste Gas Temperature	approx. 395°C
Fuel	natural gas
Net Calorific Value	31 700 kJ/m^3
2. *Compressor*	
Type	two casing, with gearbox
Speed, LP	3000/minute
Speed, HP	7622/minute
Shaft Power	approx. 60 MW
Air Inlet	1.013 bar, 10°C
Air Outlet	60 bar, 50°C
Air Flow Rate	108 kg/s
Number of Intermediate Coolers	3
Number of After-coolers	1
3. *Air Storage* (see Table 4.1)	
4. *Cooling Water*	
Type	cooling tower
Flow rate	
Discharging	1100 m^3/h
Charging	5600 m^3/h
5. *Storage Data*	
(a) *Charging*	
Compressor Power (rated)	60 MW
Duration	8 hours
Charging Energy	468 MWh
(b) *Discharging*	
Discharge Power	290 MW
Discharge Duration (full load)	2 hours
Discharge Energy	580 MWh
Fuel Energy	945 MWh
(c) *Flexibility*	
Normal Start-up Time (to full load)	11 minutes
Quick Start-up	6 minutes
Compressor Start-up Time	6 minutes
Max. Speed of Load Pick-up	88 MW/minute
(d) *Dimensionless Storage Data*	
Ratio of Charge Time to Discharge Time	$8/2 = 4$
Ratio of Discharge Energy to Charge Energy	$E_p = 580/468 = 1.24$
Ratio of Discharge Energy to Fuel Energy	$E_b = 580/945 = 0.615$
Turnaround Efficiency[1]	$\eta_{st} = 0.55$

[1] Equivalent storage efficiency [7.7]: With a thermal efficiency of a gas turbine with recuperation of 0.34, the turnaround efficiency of the storage part of the CAES is $\eta_{st} = (580 - (0.34)945)/468 = 0.55$

generator drives the compressor via a gearbox and there are three intermediate air coolers and one after-cooler. Charging requires 60 MW electric power from the grid. It takes 8 hours to increase the air pressure in the salt domes (totalling 300 000 m^3) from 45 to 65 bar. During discharge, an HP furnace chamber and a LP furnace chamber placed between the HP and LP turbines are in operation. The total electrical power is 290 MW, while the storage capacity is 580 MWh or 2 hours full load. Table 7.1 shows further data and Fig. 7.5 presents the overall arrangement of the Huntorf plant, including the geological features. A second CAES plant is being built in the United States [3.3].

Further improvements of CAES plant in terms of specific storage capacity E_r (kWh/m^3), of the ratio of peak energy to pumping energy E_p, of the ratio of peak energy to fuel energy E_b and of equivalent storage efficiency [7.7] may be achieved by one or more of the following measures:

- Regenerative heating of air by flue gas as shown in Fig. 7.6 A could improve E_b from about 0.60–0.65 to 0.8–0.85 [7.8] and storage efficiency η_{st} from 0.55 to 0.67–0.80.

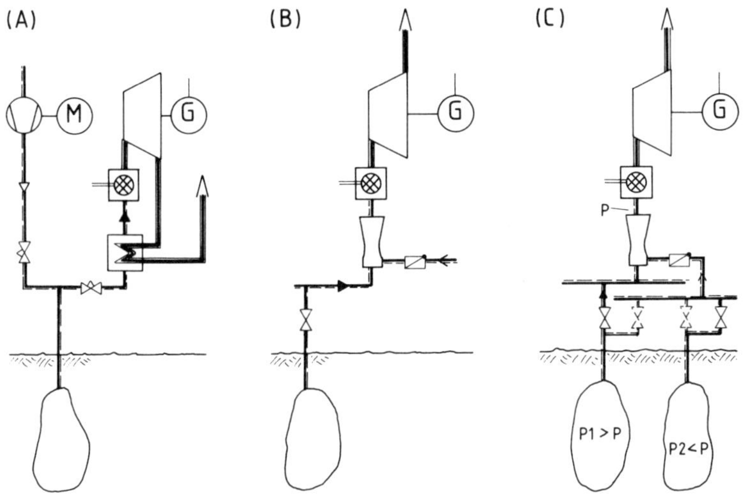

Fig. 7.6. Compressed air energy storage
A with regenerative heating of air, *B* suction side of ejector is connected to the ambient air, *C* suction side of ejector is connected to the discharge cavern

- Separate thermal energy storage could also improve E_b and η_{st} but is somewhat in contradiction to regenerative heating by flue gas.
- Ejectors may be used to minimize the throttling losses of sliding-pressure storage systems. Figs. 7.6 B and C show an ejector between storage and turbine; its suction is connected either to ambient (B) [7.9] or to a discharged cavern (C), thus reducing its discharge pressure [7.10]. Such ejectors should be designed with variable cross-sections in order that they may be adapted to the

varying pressures. Savings stemming from the ejectors may be considerable. With an ejector efficiency of only 50%, an improvement of specific storage capacity by 28% and a reduction of compressor energy by 2% are claimed [7.10].
- Coal-fired furnace chambers with fluidized bed combustion have also been proposed, thus extending the possible range of fuels to include solid fuels [7.11].

7.3.3. Air Storage for Open-cycle Gas Turbine Quick Start-up

A further application of compressed air storage in power plants is its use for the quick start-up of open-cycle turbines for reserve (stop-gap) duty [7.12]. In comparison to other thermal plants, the gas turbine may be started up rather quickly, even without air storage (minimum start-up time: about one minute). However, its lifetime is severely reduced by such quick start-ups. Therefore, the quick start-up is avoided in actual operation if at all possible and is practically eliminated. The gas turbine thus looses its stop-gap properties and stop-gap power must be found somewhere else. This is where high-pressure air storage may come in. The open-cycle gas turbine with air storage as described below is able to start up very quickly and to supply net electric power almost immediately. Contrary to the air storage system described above, relatively little storage volume is required. Therefore, above-ground compressed air storage (for example, in PCIV) may be used.

The basic idea is to increase temperature as slowly as necessary in order to avoid any marked reduction of lifetime and, at the same time, an immediate availability of the full power output (that is, the rate of power output increase is as fast as the generator can take it). It should be noted that without the storage system, power for the start-up motor must be taken from the grid in the most critical phase.

Fig. 7.7 shows the arrangement. In order to become a quick start-up set, the single-shaft open-cycle gas turbine is equipped with the following additional features:

- a start-up damper *1* before the compressor;
- a non-return valve *2* after the compressor;
- a vent valve *3* between compressor and non-return valve;
- a high-pressure air storage vessel *4*;
- a throttling valve *5* for air flow control from the storage vessel; and
- a charging set *6* for the air storage system, consisting of the (relatively small) charging compressor (which may consist of several compressors in series that can be put into operation consecutively as storage pressure rises) and a non-return valve.

The procedure is shown in Fig. 7.8 for a 100 MW gas turbine plant. Before start-up, the throttling valves *5* and the start-up damper *1* are closed and the vent *3* is open.

Start-up occurs in the following steps:

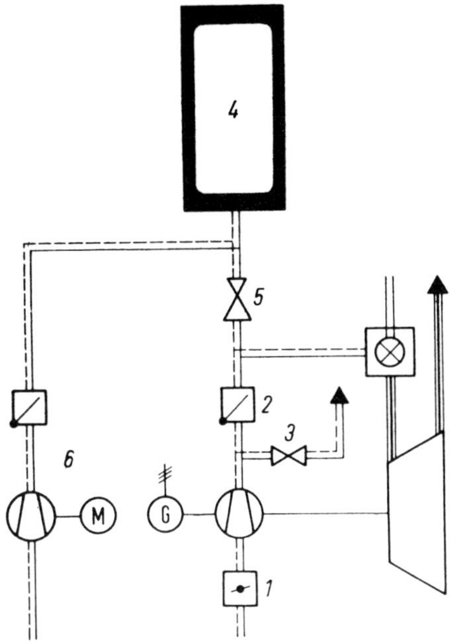

Fig. 7.7. Quick start-up of a gas turbine plant with PCIV air storage [7.12]
1 start-up damper, *2* non-return valve, *3* vent, *4* air storage PCIV, *5* throttling valve, *6* charging compressor

- The throttling valve *5* opens; the set accelerates; the fuel pipe opens; turbine inlet temperature is controlled by flow rate of fuel.
- The generator is synchronized.
- The throttling valve *5* opens further until rated or maximum power is reached; the pressure before turbine is lower than rated because the compressor is idling.
- The vent *3* closes.
- The start-up damper *1* opens slowly until the turbine inlet pressure approaches rated value; the throttling valve *5* closes slowly so as to keep power output constant.

Full turbine inlet temperature is ordinarily reached after 10 minutes; the start-up damper is fully open and the throttling valve *5* is fully closed. The combustion chamber is on full power and the start-up procedure is now completed.

In the example shown the supply pressure at the end of discharge is 8 bar. The storage pressure therefore is to be much higher. It should be noted that the air storage vessel takes much more air mass at higher pressure (see Section 2.6). Therefore, high or very high pressures should be used. Also, the vessel should have a large mass in order to limit the air temperature decrease during discharge. Both effects favour the use of the PCIV.

In the example (100 MW net power, 10 minutes), the required air mass is 150 tonnes. If a storage pressure of 150 bar is assumed, then according to the

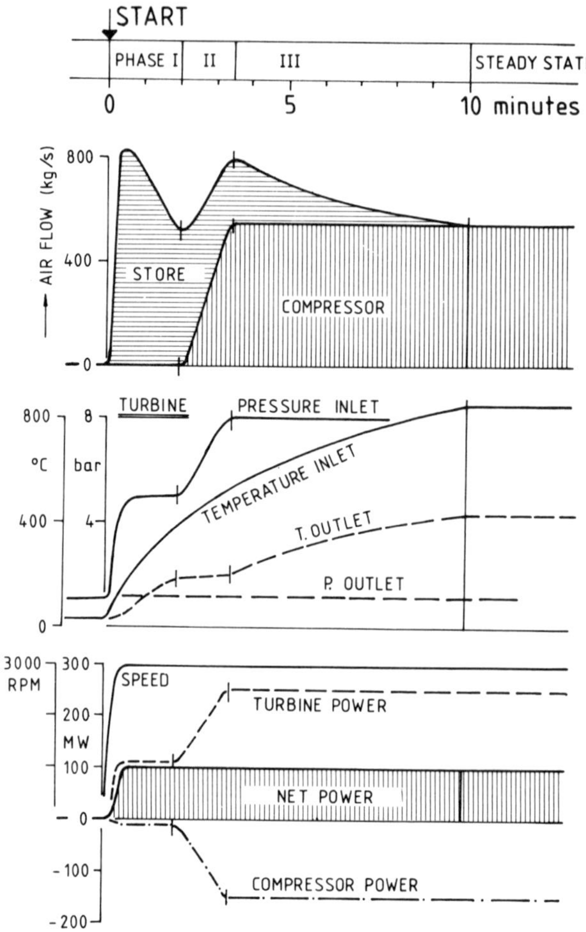

Fig. 7.8. Start-up procedure of a gas turbine plant with PCIV air storage [7.12]

diagram in Fig. 2.14, the required net pressure vessel volume becomes 150 000/164 = 920 m^3. Cost estimates show that the PCIV which requires no thermal insulation may cost (with valves, piping, charging compressor and additional control) about 20% of the gas turbine plant or correspondingly less for a combined gas/steam turbine plant. Considering the possibilities of quick start-up without lifetime reduction of the gas turbine, this should be a worthwhile expenditure.

7.4. Thermal Storage in Coal-fired Power Plants

Applications of thermal storage to coal-fired power plants date back many decades. The first purpose of thermal storage in coal-fired plants was to equalize the boiler load. Later instantaneous reserve and peak shaving came to be additional functions. Separate storage turbines were sometimes used for phase-shifting duty.

When boilers were subsequently equipped with easily controlled pulverized coal firing and therefore became more flexible (and, at the same time, large electric grids were constructed), there was less need of thermal storage. Only relatively small storage plants remained to be built for control purposes and for start-up [7.13].

Renewed interest in thermal storage in coal-fired power plants stems from

- the need to save valuable fuel such as oil or gas which were heretofore used for peak and medium load and
- the need for better utilization of coal-fired plants since such facilities have become more and more complex and costly because of expensive environmental control equipment such as electrostatic precipitators and flue gas desulphurization plants.

Suitable for coal-fired power plants are: sliding-pressure storage without or with superheat storage, direct or indirect feed water storage and cascading storage.

7.4.1. Built Peak and Instantaneous Reserve Plants

(a) *Sliding-pressure Storage*

The first power plant application of sliding-pressure storage (steam storage) was in the power station of Malmö, Sweden. This was an instantaneous reserve plant of 3.75 MW, with storage pressure of from 8 to 2 bar, bridging the gap of 50 minutes between the load shedding of a hydro plant and the take-over by a thermal power plant after cold start-up.

Subsequently, a number of power stations for electric railways, with peak loads at the rush hours, were equipped with sliding-pressure storage. Some typical data are: a few MW power, a few hundred m^3 storage volume and upper storage pressures of up to 17 bar.

The ultimate in developments of this type and, at the same time, the largest plant built is the steam storage power plant in Berlin-Charlottenburg (see Chapter 1, Figs. 1.22 and 1.23). Fig. 7.9 shows the flow sheet of this plant that was built in 1929 and is still in operation today. Steam from topping turbines charges the 16 vertical storage vessels of 312.5 m^3 each (5000 m^3 total). The upper storage pressure is 14 bar, while the lower pressure is 1.5 bar. During peaks, steam from the store is admitted to two storage turbines of 20 (25) MW electric power each. Full load capacity (referring to a lower storage pressure of 4.5 bar) is 1.08 hours or 1.08(40) = 43 MWh. The total capacity (down to 1.5 bar) is 67 MWh, yielding a volumetric storage capacity of 67 000/5000 = 13.4 kWh/m^3.

Fig. 7.10 shows one of the turbines. It has two sets of intake control valves which are necessitated by the large pressure swing. In order to minimize idling losses, the turbine has a low number of stages and is rather short.

High pressure steam storage, which was first used in locomotives (see Chapter 8), has also been applied to power stations. Its advantages are: no need for

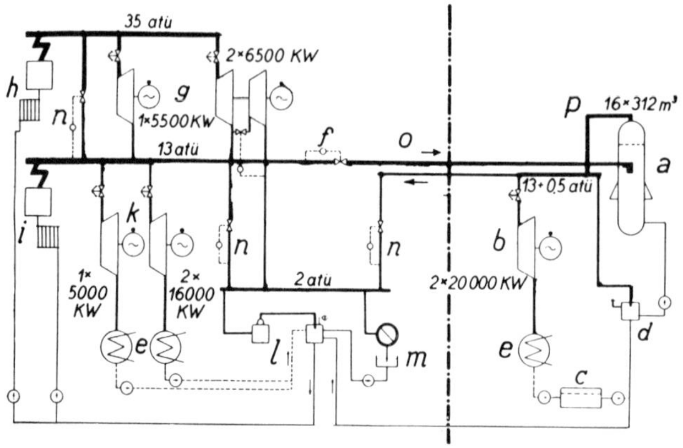

Fig. 7.9. Flow sheet of the steam storage plant at Berlin-Charlottenburg [7.14]
a sliding-pressure (Ruths) storage, *b* storage turbine, *c* condensate store, *d* degasser for storage make-up, *e* main condenser, *f* pressure control valve, *g* topping turbine, *h* HP boiler plant, *i* LP boiler plant, *k* LP turbines, *l* make-up water, *m* district heating grid, *n* pressure reduction valve, *o* charge line, *p* discharge line

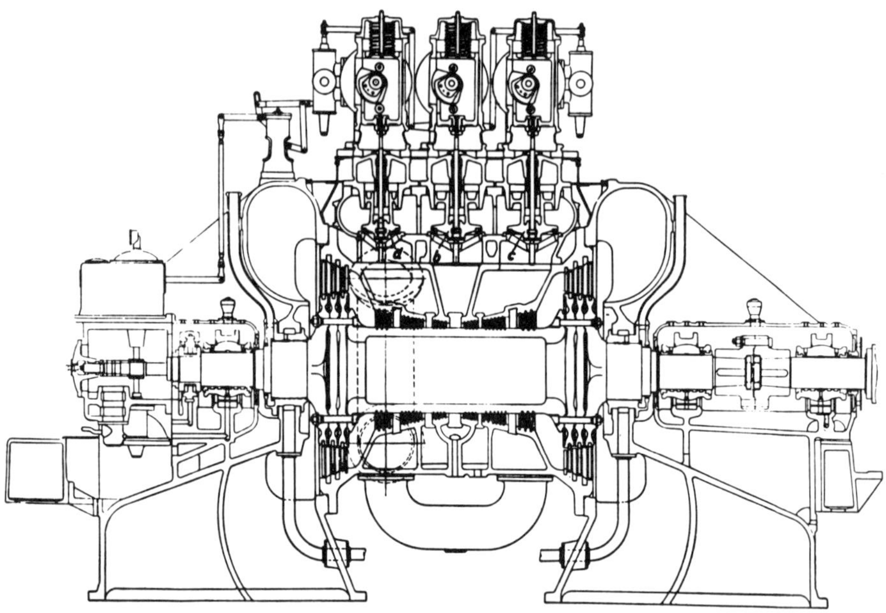

Fig. 7.10. Multiple admission turbine ("storage turbine"), 20/25 MW [7.14]

special storage turbines because of constant pressure discharge, high volumetric storage capacity (if vessels with hot wall are used, 25 to 35% of the capacity is stored in the vessel wall), low space requirement and low insulation losses.

In 1938, an instantaneous reserve plant of this type was operational at the Simmering power station of the Vienna municipal utility. The plant served to bridge the 15 minutes gap until a coal-fired forced-circulation boiler of 100 tonnes/hour steam of 40 bar, 425 to 450°C was started up from cold by oil firing [7.16].

Fig. 7.11 shows the flow sheet. The upper pressure was 120 bar, while the lower pressure was normal turbine pressure of 32 bar. There were six high-pressure drums of 9.25 m length and 1.25 m inner diameter. The wall thickness was 72 mm, while that of the flat ends was 300 mm. The total volume was 63 m^3. The four lower drums were completely filled with water, while the two upper ones had a steam cushion.

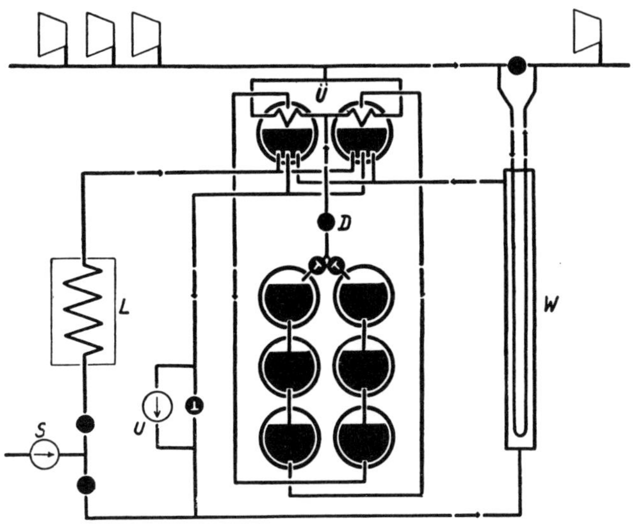

Fig. 7.11. The high pressure storage plant at Vienna, built in 1938 [7.15]
D throttle valve, *L* charging boiler, *S* feed pump, *U* recirculation pump, *W* charging heat exchanger, *Ü* superheater

There were also two 120 bar superheat storage vessels which had a total volume of 19 m^3. Charging was done by a small boiler or by a heat exchanger partly desuperheating a live steam by-pass ("steam transformer"). The storage capacity was 4 MWh and the volumetric capacity 49 kWh/m^3. A picture of the vessels, without thermal insulation, is shown as Fig. 1.25.

(b) *Displacement (Constant Pressure) Storage*

In steam power plants (including cogeneration plants) with regenerative feed water heating by bleed steam, the displacement (constant pressure) storage system with vertical vessels gained acceptance during and after the late 1920s. This is a true flow storage system (see Chapter 3). Therefore, the turnaround efficiency is rather high as is volumetric storage capacity.

The storage plant of the coal-fired Mannheim power station is well known [7.17]. The power of this plant was 21.15 MW without storage and 25.4 MW (+ 20%) during discharge of the displacement store. The storage vessel had 3 m diameter, 20 m height, 135 m³ volume, 20.5 bar operating pressure, 250/70–80°C operating temperature, a capacity of 5 MWh and a volumetric capacity of 37 kWh/m³.

Charging of the storage vessel permitted a reduction of the electric load down to 12.25 MW. This means that the ratio of charge power to discharge power was 8.9/4.25 = 2.1 and is much higher than for pumped storage. The basic flow sheet is shown in Fig. 7.12 [7.21]. Control of the storage system was achieved by means of the control valve *R*. If *R* is open, excess feed water is stored (charging); if *R* is closed, then all of the feed water is displaced from the storage vessel by cold water (discharge). The water level is controlled by the cold water supply control *W*.

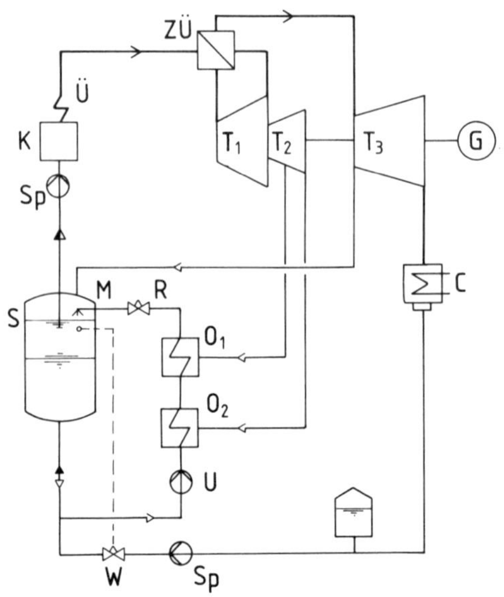

Fig. 7.12. The Mannheim feed water storage plant [7.17]
K boiler, *Ü* superheater, T_1 topping turbine, T_2 regenerative turbine, T_3 MP condensing turbine, *C* condenser, O_1, O_2 feed water heaters, *Zü* reheater, *S* storage vessel, *M* charging equipment (mixing preheater), *R* control valve, *W* water level control valve, *Sp* feed pump, *U* recirculation pump

(c) *Buffer Storage*

The purpose of buffer storage in steam power plants is to increase the amount and, in particular, the speed of load changes [7.13]. Fig. 3.18 shows a sliding-pressure buffer storage vessel which is connected to the cold reheat line (see Section 3.4.3). Such buffer storage vessels were built in standardized power plants in the 1940s.

Fig. 7.13 shows an expansion-type buffer storage in the feed water line [3.6]. Hot water is extracted from the storage vessel and then flashed in several flash evaporators (two, in the case shown) and the steam is then admitted to feed

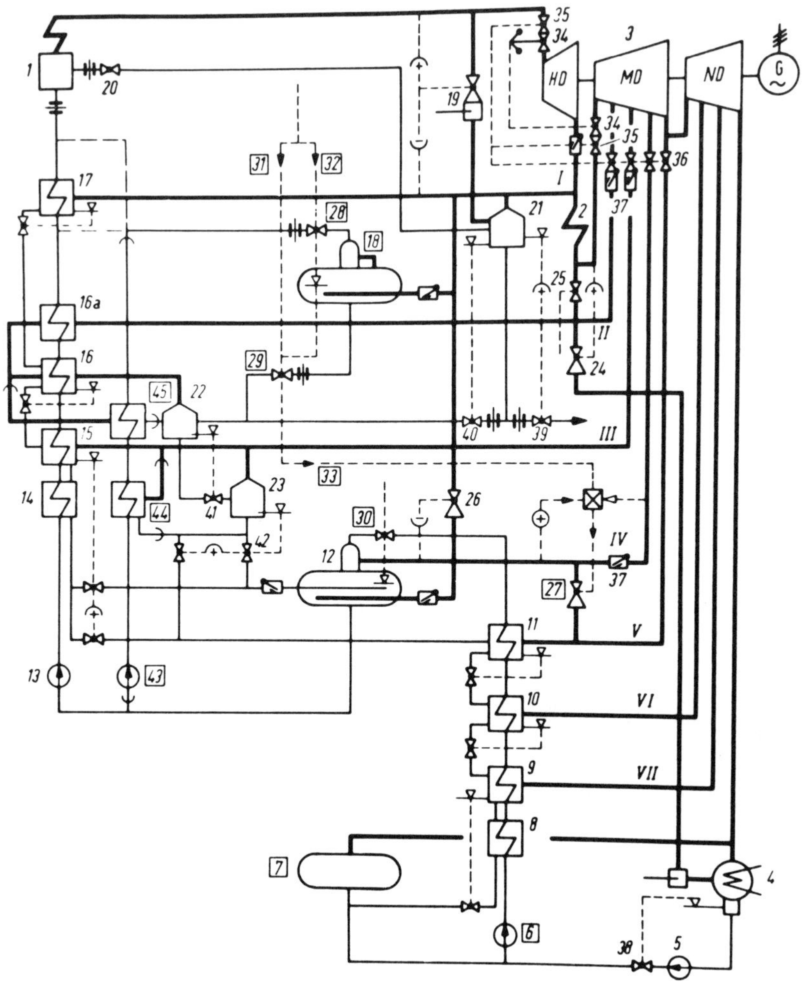

Fig. 7.13. Integrated feed water storage [3.6]
1 boiler, *2* reheater, *3* HP, MP, LP turbine, *4* condenser, *5, 6* condensate pumps, *7* cold store, *8* condensate cooler, *9, 10, 11* LP feed water heaters, *12* feed water storage tank with mixing condenser and degasser, *13* feed pump, *14* condensate cooler, *15, 16, 17* feed water heaters, *16a* desuperheater, *18* hot water store with degasser, *19* live steam by-pass and desuperheater, *20* low load by-pass, *21, 22, 23* flash evaporators, *24, 25* MP by-pass and desuperheater, *26, 27* by-pass, *28* store-filling valve, *29* discharge valve, *30* water level control, *31, 32, 33* storage control signals, *34* turbine control valve, *35* fast acting turbine control valve, *36* fast acting bleed control valves, *37* bleed non-return valves, *38* to *42* water discharge valves, *43* storage feed pumps, *44, 45* storage feed heater, *I* to *VII* bleed and feed water heater stages

water heaters and, if required, even to the turbine. The feed water train is used for charging the storage vessel. Control is designed in such a way that during normal operation of the power plant there is neither charge nor discharge and that there is a medium water level in the storage vessel. Only in case of fast load changes use is made of the storage system.

Start-up storage may be regarded as a special case of buffer storage. Excess steam produced after the start-up of the boiler but before the start-up of the turbine and the generator (and in case of load shedding) is stored in order to save heat and condensate.

7.4.2. New Proposals for TES in Coal-fired Power Plants

A renewed interest in TES for coal-fired power plants, an interest that is mainly directed to peaking purposes, has recently developed.

(a) *Underground Sliding-pressure Storage*

Fig. 7.14 shows a peaking system with rock caverns (see Section 4.4), while Fig. 7.15 [7.18] presents the flow sheet of the peaking system. Five special vessels having a height of 38 m are envisaged. These will act as sliding-pressure store between 48 bar (715 psi; HP turbine outlet) and 24 bar (355 psi; constant peaking turbine inlet pressure). In spite of throttling losses during charging and discharging, the turnaround efficiency is about 80%.

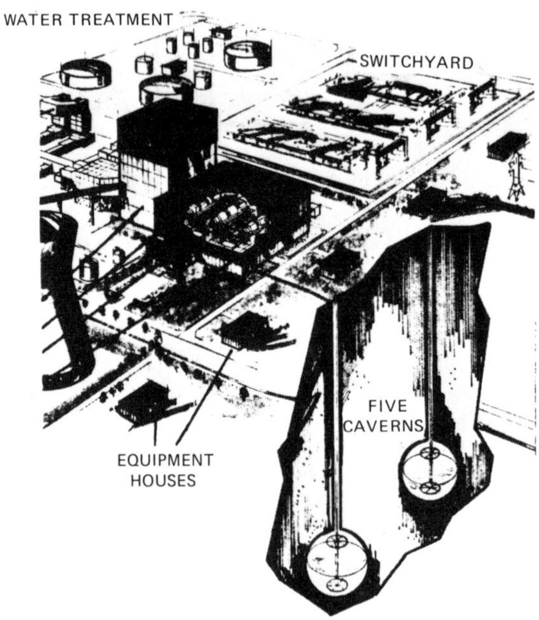

Fig. 7.14. Sliding-pressure storage plant with underground vessels [7.18]

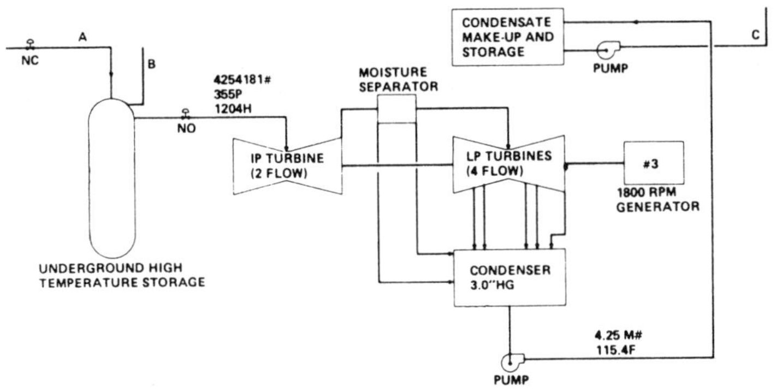

Fig. 7.15. Flow sheet of plant according to Fig. 7.14 [7.18]

(b) *Indirect Feed Water Storage*

The economy of a coal-fired power station with indirect feed water storage by means of hot oil as storage medium was investigated by Bechtel [7.19] for the retrofitted case. The flow sheet is shown as Fig. 7.16. The storage system allows an increase of the power from 595 MW to 571 + 120 = 691 MW (that is, by 16%)

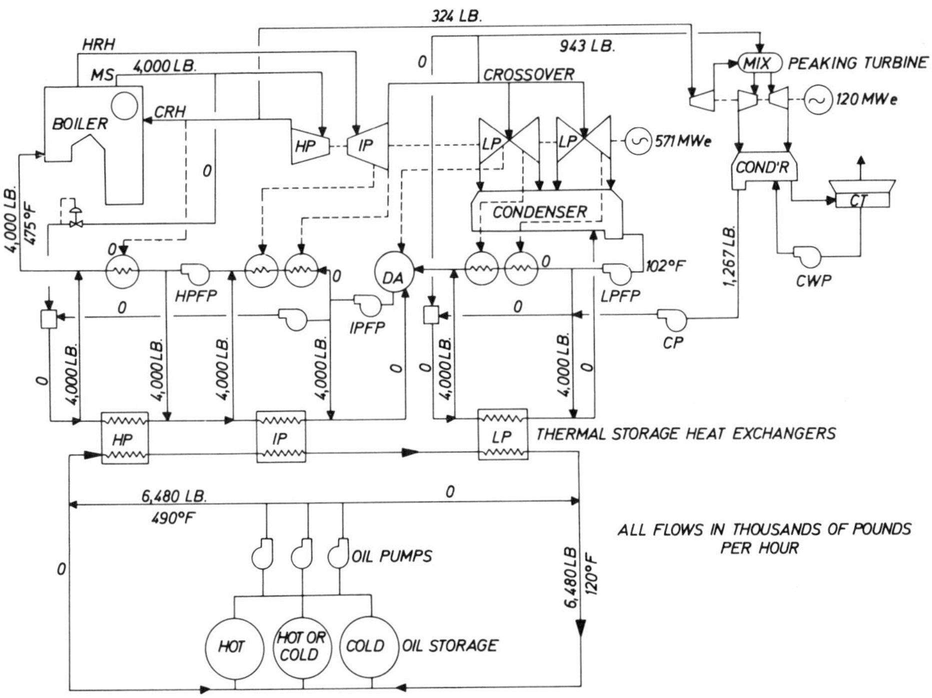

Fig. 7.16. Coal-fired power station with a retrofitted indirect feed water storage–conditions during discharge [7.19]

during 8 hours. There are water/oil heat exchangers in parallel to the whole feed water heater train. The tank farm consists of three tanks of 36.6 m (120 ft.) in diameter and 18.3 m (60 ft.) in height, containing rocks. Two of the tanks are oil-filled in the charged condition, the other one is empty (filled with inert gas). Exxon Caloria HT-43 is proposed as the heat transfer and storage medium.

Feed water is preheated in the oil/water heat exchangers during discharge, thus permitting an almost complete closing of the bleed lines. Therefore, exhaust steam is increased by almost 50%. An additional peaking turbine is suggested, since the main turbine will not normally be able to handle this increased flow rate. Charging is done by heating additional feed water and transmitting its heat via the heat exchangers to the oil. Turnaround efficiency is 61%. The study concludes that at least retrofitting of the TES will be uneconomic, because it requires that the plant be shut down for 16 months.

A similar system (see Fig. 7.17) was investigated by General Electric [7.18] for the grassroot case of a 741 MW base load plus 400 MW peak load plant. A supercritical steam process without reheat is proposed. Heat exchange is now between MP steam and oil, permitting a higher peak load (54% increase). Turnaround efficiency is calculated as 66%.

(c) *Direct Feed Water Storage*

Peaking by direct feed water storage was also proposed. It requires pressure vessels instead of tanks, but in comparison with the indirect system it saves heat exchangers, their exergy losses and oil.

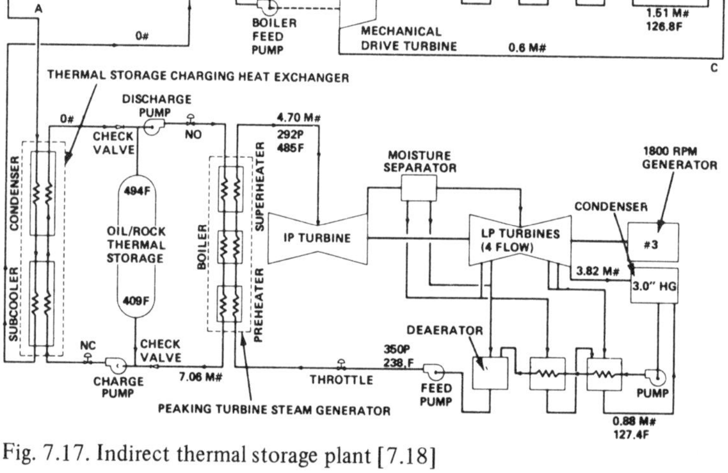

Fig. 7.17. Indirect thermal storage plant [7.18]

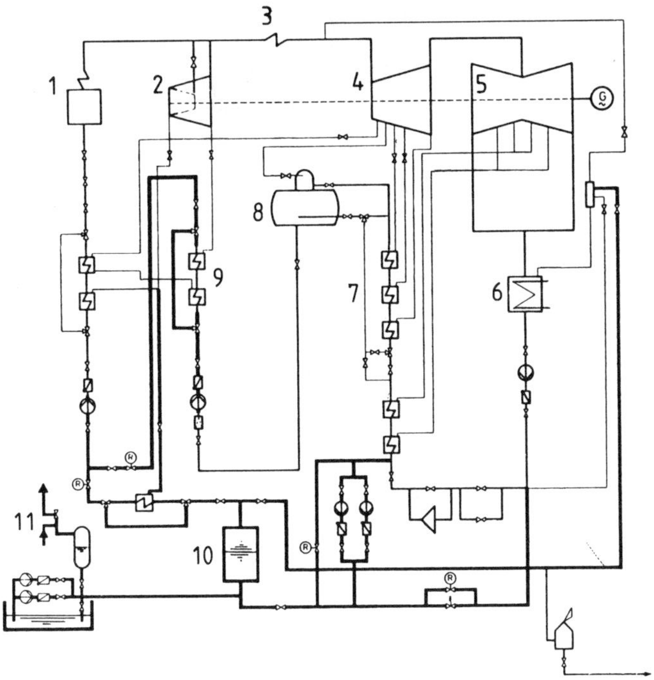

Fig. 7.18. PCIV feed water storage for a 150 MW plant [4.24]
1 boiler, *2* HP turbine, *3* reheater, *4* MP turbine, *5* LP turbine, *6* condenser, *7* LP feed water train, *8* degasser, *9* HP feed water train, *10* pressurized hot water storage, *11* pressurizer

Fig. 7.18 shows a proposal for an existing 150 MW plant, increasing its power by 13.5% during 3 hours [4.24]. The prestressed cast iron vessel (PCIV, see Section 4.5.3) with the design data 70 bar, 242/47°C, 2000 m^3 is of the cold-liner type, that is, it has thermal insulation on the inside. The turnaround efficiency is said to be 95%. Economic advantages are claimed for this system versus a peaking gas turbine and versus hydraulic pumped storage if 3 to 5 hours discharge per day and 150 to 200 full discharges per year are assumed.

(d) *Combined Feed Water and Steam Storage (Cascading)*

In order to improve the part load behaviour and, at the same time, increase the peak load above the narrow limits applying to the pure feed water storage systems (see Section 3.3.3), the combined feed water and steam storage (cascading) system (Fig. 7.19) was proposed [4.23, 7.23, 7.25].

The saturation temperature in the storage vessel is now above the final feed water temperature. During charging, the feed water is therefore to be heated up to storage temperature (for example, by mixing with live steam or reheated steam) in the upper part of the vessel. There will be throttling losses. But, on the other hand, the advantage of constant storage temperature is gained, whereas with pure feed water storage this would require additional heaters.

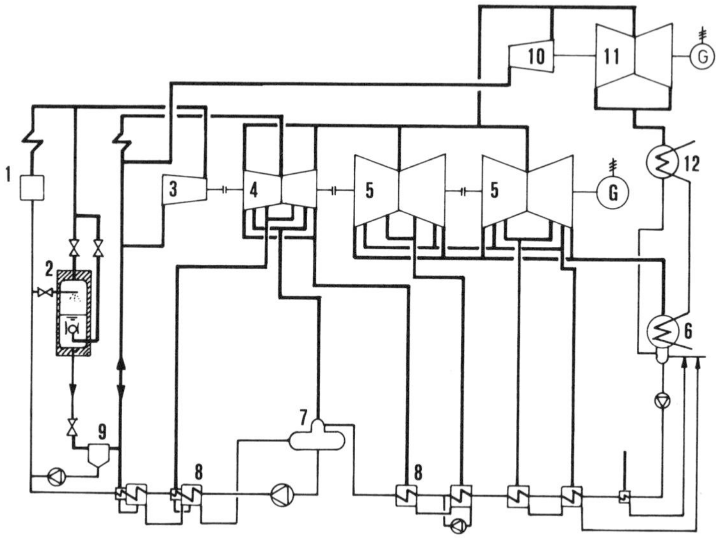

Fig. 7.19. Basic flow sheet for cascading (feed water and steam) storage [4.23]
1 heat source, *2* expansion-type PCIV, *3* HP main turbine, *4* IP main turbine, *5* LP main turbine, *6* main condenser, *7* feed water tank (cold store), *8* feed water heater, *9* flash evaporator, *10* HP peak load turbine, *11* LP peak load turbine, *12* peak load condenser

During discharge, hot water is (according to the principle of expansion storage) taken from the vessel and flashed to the pressure of the highest bleed. The steam is piped to the appropriate turbine stage and the hot water replaces (partly or fully) the feed water. There are no feed water mixing losses as the water temperature from the flash evaporator and from the feed water heater train always has the same temperature.

A PCIV is proposed as pressure vessel. Since there are (contrary to the situation in displacement storage and sliding-pressure storage) no appreciable thermal gradients, the PCIV may be designed with cold or hot walls, that is, with internal or exterior thermal insulation. The cost of the PCIV increases less than proportionally with pressure and, in this way, makes the high storage pressures and thus peak loads feasible. Accordingly, peak loads that are largely in excess of those possible with pure feed water storage are also thus made possible (see Section 3.3.3).

7.5. Thermal Storage in Nuclear Power Plants

There is even more incentive to use energy storage for nuclear power plants than there is with coal-fired plants. This is so for the following reasons:

- Energy storage permits to go away not only from oil but from fossil fuel altogether. (In the case of a coal-fired plant, this switch, however important,

is merely from one fossil fuel to another.)
- Nuclear fuel is relatively inexpensive.
- The initial cost of the reactor island is very high and energy storage permits maximum use of this capital cost intensive item.
- Cyclic operation of a nuclear reactor is detrimental to its fuel elements. Base load operation of the reactor island minimizes temperature cycling of the fuel.

Energy storage is thus the safest and most economical way of reactor operation [7.20, 7.21]. Use of the nuclear power plant for peaking by means of energy storage also reduces the need for separate peak load plants as well as the need to use scarce and expensive peak load fuel (light oil, kerosene and natural gas) with low efficiency.

Integrated thermal energy storage within the nuclear power plants itself is one of the most promising possibilities of energy storage. However, there are limitations:

- For reasons of radioactivity, it will be difficult to store thermal energy in the primary cycle.
- The turbine and generator may be at their limit capacity. This applies to Light Water Reactors which are now built in units of 1200 to 1300 MW; it does not apply to certain other types of reactor such as the Canadian Heavy Water Reactor CANDU which is presently built in units of about 600 MW. Systems with overloadable turbine are therefore possible for some (but not all) reactor types, the others requiring a separate peaking turbine for integrated TES in any case.
- Even if the overloading of the main turbine set was theoretically possible, there may nevertheless be a tendency to separate base load set and peaking equipment as much as possible. However, the connection between base load set and peaking station may be reduced to a few pipe lines for charging the storage vessels and returning condensate.

Several systems of integrated TES in nuclear power plants have already been suggested since the 1950s:

- Smidt has pointed to the basic economic advantages in 1958 [7.20].
- Marguerre has proposed feed water displacement storage in spherical vessels [7.21].
- Margen [7.22, 4.10] and later Dooley [4.8] have proposed TES in rock caverns.
- Pacault has proposed a steam pumped storage system with direct drive of the steam compressor [3.5].
- Gilli and Beckmann [2.9] proposed the expansion-type storage system.
- Bechtel [7.19] and Hall [7.18, 1.25] discussed indirect TES in Thermocline rock and oil storage tanks.
- Displacement storage with external steam generation by flashing was proposed for use with the CANDU reactor [4.11].
- Gilli and Beckmann [4.23, 7.25] and Oplatka [7.23] proposed the cascading system with steam/feed water storage.

The main systems are described in the following.

7.5.1. Feed Water Storage

Marguerre [7.21] had suggested the pure feed water system as had Margen [7.22]. However, the latter had restricted it to temperatures above 73°C (that is without using the lowest two feed water heaters). There may be good reasons for doing so, but it reduces the possible peak load.

Hall [1.25] and others [7.18] have also discussed feed water storage in PCIV at 40 bar (600 psi) for Light Water Reactor power plants.

7.5.2. Indirect Feed Water Storage

The indirect system with heat transfer oil (Exxon Caloria HT-43) and rocks proposed by Bechtel for coal-fired plants [7.19] was discussed by Hall [1.25] and others [7.18] for nuclear plants.

7.5.3. Steam Storage Systems

Feed water systems are rather limited in peak power and affect the main set considerably. By contrast, steam storage systems are theoretically unlimited in peak power and are really separate plants connected to the main plant by only a charging line and a condensate return. This may be considered an advantage for application to nuclear plants.

The main systems of this type are (see Chapter 3): internal steam generation (Ruths type) and external steam generation (expansion type or displacement type).

Pure sliding-pressure (Ruths) type storage was proposed for rock cavern storage [4.8]. The flow sheet of a Ruths type storage plant with sequential admission and some superheat is shown in Fig. 7.20 [7.24].

Of the systems with external steam generation, multiple flashing has been suggested as well as displacement (Thermocline) storage as shown in Fig. 3.15 [4.11].

7.5.4. Overloadable Turbine versus Separate Peaking Turbine

If the reduction of the bleed flow is considered unsuitable for a nuclear plant, or if the overloading of the turbine set is not possible, then a separate peak load turbine is required. The arrangement may be similar to the one of the coal-fired station shown in Fig. 7.19.

Advantages of the separate peaking turbine include the removal of power limitations caused by either the feed water storage and/or the admission limit of the main turbine. The fact that TES with the main set constitutes a spinning reserve may be considered an advantage of this system. However, idling of the peaking turbine is also possible (although with some loss of energy).

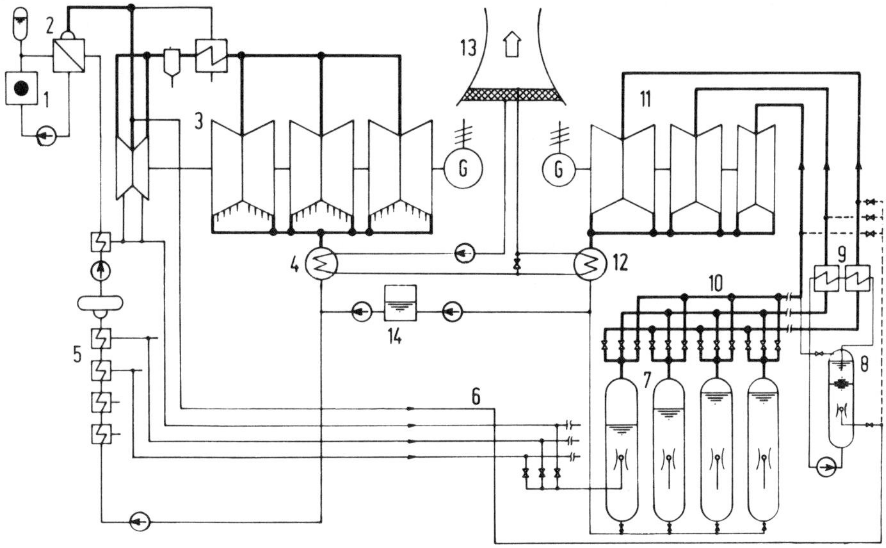

Fig. 7.20. Flow sheet with sliding-pressure storage, sequential discharge and separate peak load turbine [7.24]
1 reactor (PWR), *2* steam generator, *3* main (base load) turbine, *4* main condenser, *5* feed heater train, *6* charge line, *7* main accumulators, *8* superheat accumulator, *9* superheater, *10* discharge steam lines, *11* peak load turbine, *12* peak load condenser, *13* cooling tower, *14* condensate storage

7.5.5. Combined Feed Water and Steam Storage (Cascading)

In order to exceed the power restrictions of pure feed water storage, Marguerre [7.17] had already suggested a series of external steam generators producing LP steam from the heat of the feed water storage vessels and applying it via the bleed lines to the main turbine. In a modification (Fig. 7.21), he proposed to use the feed water heater train (with the exception of the highest heater) as steam generator train in a two-fold manner: Steam generation takes place via the heat transfer surface of the heaters by means of pumping stored hot feed water back to and through the heater and by simultaneous flashing on the steam side. In any case, not only the bleed flows are reduced or eliminated, but there is even steam flow back through the bleed lines (except the highest one) to the turbine stages. However, this may be limited by the admission capability of main turbine and the main condenser as well as by the need for stage moisture control. This is particularly so in the saturated steam turbines of nuclear power plants. Alternatively, a separate peaking turbine will have to be used. An increase of possible discharge power, compared to the pure feed water system, by a factor of about 2 appears possible under certain circumstances.

Another possibility for increasing peak power capability which retains the advantages of feed water storage is the combination of feed water storage with a flash evaporator as topping process (see Section 3.1.2). Fig. 7.22 gives an example

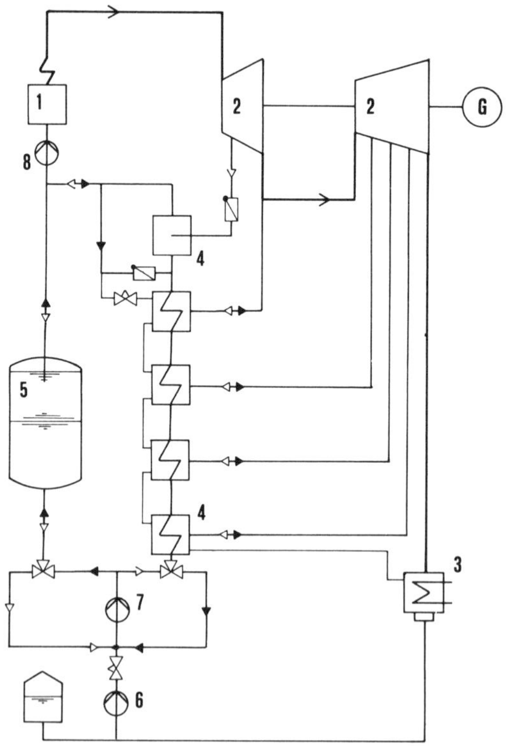

Fig. 7.21. Combined feed water and steam storage as proposed by Marguerre [7.21]
1 steam generator, *2* turbine, *3* condenser, *4* feed water/steam generator train, *5* storage vessel, *6* condensate pump, *7* store recirculation pumps, *8* feed pump

of such a cascading flow sheet, applied to a PWR with overloadable turbine set [7.25]. In this case, the only additions to the base load plant are the PCIV, a flash evaporator, an enlargement of the feed water tank (or a separate cold storage tank) and some piping. The charging of the PCIV is performed by means of live steam plus hot feed water being mixed in the PCIV. The water content of the PCIV is flashed during discharge. The steam generated in this way is piped to the highest extraction point, possibly after some superheating. However, part of this steam is used to finally heat up some feed water that cannot be replaced by water from the PCIV. This is because there is moisture from the moisture separator and hot condensate from the reheater that has to be used. There is also some feed water that must be heated up by the minimum amount of extraction from lower bleed points that is necessary for reason of moisture control at the cold end of the turbine.

Instead of an expansion-type storage system, a displacement (Thermocline) system with external steam generation may be used in a cascading arrangement. There is little or no need for cold store, but the PCIV needs internal thermal insulation. Moreover, there is some pumping power.

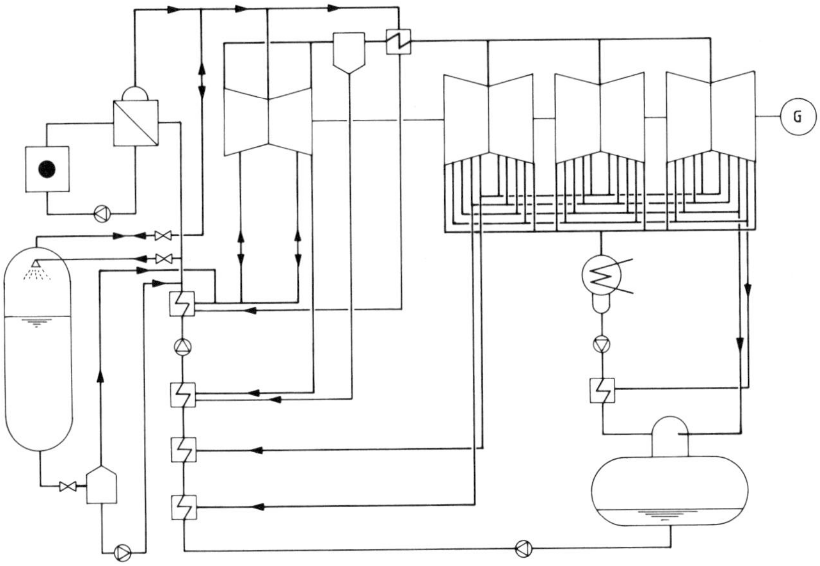

Fig. 7.22. Flow sheet of a nuclear power plant (PWR) with integrated thermal energy storage of the cascading type and turbine set capable of overload [7.25]

7.5.6. Integrated Steam Pumped Storage

According to the proposal of Pacault (Section 3.1.7, Fig. 3.8), there is a steam compressor coupled to the main set, pumping live steam from 64 to 200 bar and, at the same time, raising its temperature from saturation (300°C) to 450°C. Its heat is then transferred to hot air of atmospheric pressure and then finally to a solid body TES consisting of bricks. Discharge takes place via a reverse flow of compressed air to the slightly superheated MP steam, thus further superheating it from below 300°C to 350°C.

A possible power swing from 752 to 1000 MW is claimed. Turnaround efficiency will be limited, considering the fourfold heat transfer, the compressor losses and the blower losses.

7.6. Thermal Energy Storage in Solar Power Plants

7.6.1. General

Compared to fuelled thermal power plants, the solar thermal power plant exhibits several special features that make integrated thermal energy storage necessary [7.26].

(a) Fuel storage is not possible. Fuel control would be possible (for instance, by defocussing the heliostats) but—contrary to the fossil-fired plant—entails a loss of energy and is therefore avoided.

(b) Similar to the conditions of wind, tidal and hydro plants, the input of primary energy keeps changing in a diurnal and annual pattern and, in addition, depends on weather conditions. This means that not only is there a demand pattern but a supply pattern also. The need for energy storage is generally increased except in the rare case when the two patterns should happen to coincide. Discharge of the thermal storage may not only be actuated by a demand peak but also by a supply deficit (low or zero insolation) or by both, and charge not only by low demand but also by excess supply (excess insolation) or both. Therefore, TES will serve not only for peaking but also for bottoming.

(c) In many practical cases not only the power input but the heat sink system as well will show a cyclic or stochastic behaviour. In particular, this will apply to air-cooled plants (cooling tower operation of a steam plant). This will generally be a disadvantage for a thermal power plant in which peaks will occur during the day (that is, at high ambient air temperatures). However, it will be an advantage for the solar plant having thermal storage if part of the power is shifted towards evening and night (that is, to lower air temperature and thus higher efficiency).

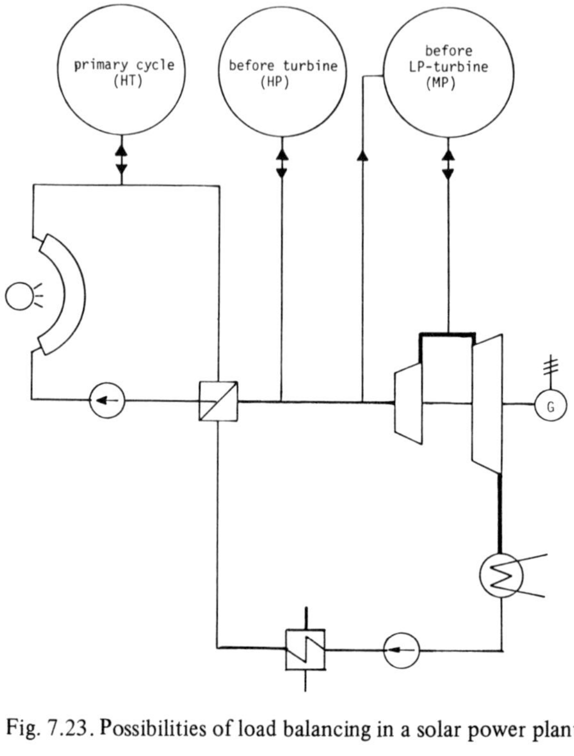

Fig. 7.23. Possibilities of load balancing in a solar power plant

(d) A short-time energy storage system becomes necessary in many climatic zones in order to cope with the effects on the plant control of sudden meteorological changes such as clouds passing by. A storage capacity of about half an hour full-load is typical for such cases.

(e) An additional primary heat transfer cycle is employed in some types of solar power plant. The primary medium may be heat transfer oil, molten salt (HITEC) or liquid alkali metal (Na). These all are suitable as high temperature storage media, too. The possibilities of integrated thermal energy storage in double-cycle plants are shown in Fig. 7.23.

(f) Feed water storage is not applicable to solar steam power plants—at least not as the only storage system. With only feed water storage during periods without solar radiation, the steam turbine would not receive steam but would receive preheated water instead.

(g) The load factor of solar plants without thermal energy storage is rather low (about 3000 h/a in arid climates, about 2000 h/a in Southwestern United States and in Mediterranean climates and about 1000 h/a under Central European conditions). By means of thermal energy storage, the load factor of all components of the plants located after the storage system will be increased. Therefore, the layout of heliostat or collector field, turbine and storage unit presents an interesting optimization problem [7.26].

7.6.2. Single-cycle Solar Steam Plants with Direct Thermal Storage

A single-cycle solar steam plant with direct thermal storage is shown in

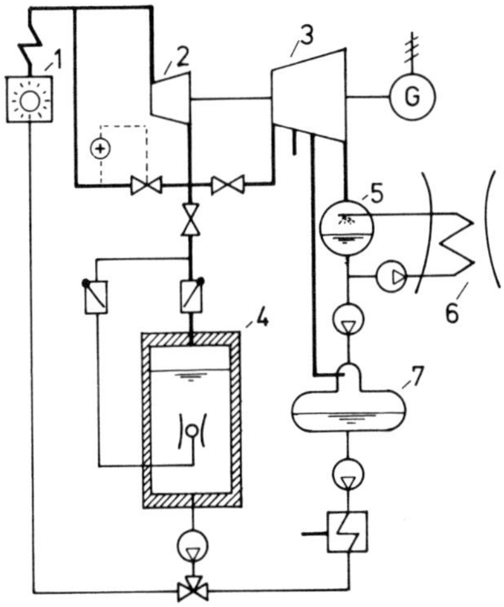

Fig. 7.24. Solar power station with PCIV sliding-pressure storage for instantaneous reserve, acting alternatively or simultaneously as an expansion-type feed water storage [7.26]
1 solar steam generator, *2* HP turbine, *3* LP turbine, *4* PCIV accumulator, *5* mixing condenser, *6* dry cooling tower, *7* feed water heater

Fig. 7.24. There are branch connections from the live steam line and from the cross-over line of the solar power plant to the storage plant which consists of a sliding-pressure type storage vessel. A mixing condenser and a dry cooling tower are shown in the example. Additional flexibility is gained by a feed water connection from the storage vessel to the feed water line [7.26].

Three typical cases of the operation of such a plant are schematically shown in Figs. 7.25a, b and c to demonstrate the principles of layout and optimization

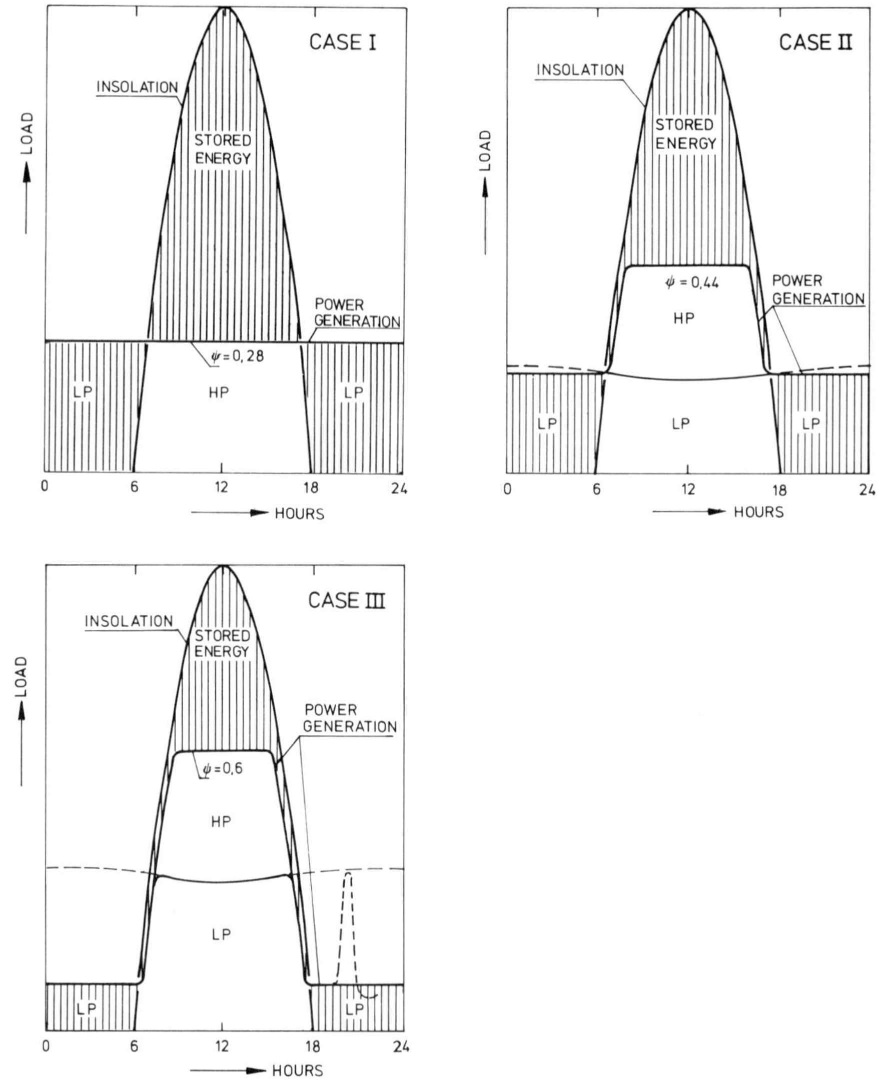

Fig. 7.25. Modes of operation and layout [7.26]
Case I: $\psi = 0.28$, Case II: $\psi = 0.44$, Case III: $\psi = 0.6$

of steam generator, storage and turbine sizes. It is to be noted that only the steam generator (but not HP + LP turbine and condenser/cooling tower) will be designed for peak insolation. An important design parameter is the "turbine layout factor" ψ which is defined as the ratio of the maximum turbine power chosen to the peak turbine power that was possible at full steam generator power (peak insolation).

Fig. 7.25a (Case I, $\psi = 0.28$) is the base-load case (bottoming case): Constant output may be achieved 24 hours a day. During the night, the low pressure (LP) turbine takes the load and all the steam is taken from the steam storage plant. As the sun rises, the high pressure (HP) part of the turbine is started up with high-pressure steam from the solar steam generator. The steam is subsequently used for charging the storage vessel. The LP inlet valve is gradually closed, thereby reducing power output of the LP turbine. The reverse action is taken in the afternoon. The shaded areas determine the necessary storage capacity. The LP output is somewhat smaller than the area "Stored Energy", because the storage efficiency is less than 1.0 (approximately 0.7 to 0.75).

Fig. 7.25b (Case II, $\psi = 0.44$) shows a somewhat different situation. The power output shape is similar to the usual daily demand pattern (there is a larger load during the day and a lower one during the night). Contrary to the operation presented in Fig. 7.25a, the LP turbine is now working all the day although its power is somewhat reduced around noon in accordance with the higher condenser pressure. The HP turbine is operated in a manner similar to the case presented in Fig. 7.25a, with the exception that now only part of the steam is used for charging the steam storage vessel, the remainder serving as working steam for the LP part. The LP inlet valve and the storage line inlet valve are now used to control the LP output on the one hand and the charging operation on the other.

In Fig. 7.25c (Case III, $\psi = 0.6$), the turbine power is increased even more than in Case II and the storage volume is correspondingly decreased. The mean value of LP turbine power during the night is therefore well below the maximum LP power (which, again, is somewhat dependent on ambient temperature). However, power peaks of relatively short duration may be accommodated up to the maximum LP power, as is shown in Fig. 7.25c by the dotted line for the evening period around 8 p.m. (for example, the TV peak).

Case IV is not specifically shown in Fig. 7.25, but would be the limiting case with $\psi = 1.0$ (that is, without any storage).

Fig. 7.26 schematically shows first cost versus the turbine layout factor ψ. The cost of the turbine, including a dry cooling tower, shows a distinct minimum for Case II. That is quite obvious, since this is the only case in which the expensive LP turbine plus cooling tower is utilized 24 hours a day. Cost of the energy storage system by definition is zero at $\psi = 1.0$. It increases almost linearly with $1 - \psi$. The required capacity of storage vessels (for instance, for Case II) is 260 MWh for 44 MW maximum turbine output. Cost of turbines plus storage is also plotted. It is lowest at $\psi = 1.0$ and increases with decreasing ψ, slowly at the beginning and more steeply at low values of ψ. For the above relation of storage cost to turbines cost, initial cost is therefore lowest without energy storage, as was to be expected.

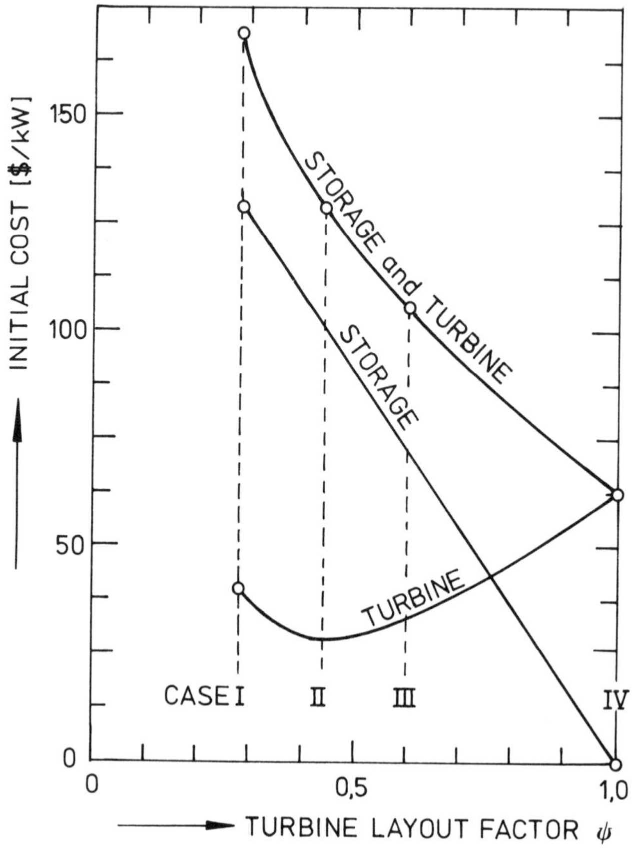

Fig. 7.26. Initial cost of turbine and thermal energy storage versus turbine layout factor ψ [7.26]

Results are different if the higher value of stored peaking energy is accounted for. The optimum layout is strongly dependent on the tariff rates and, in particular, on the ratio of the credit for stored to the credit for directly supplied electrical energy.

Fig. 7.27 shows a flow sheet with saturated steam storage of the expansion type in a solar steam plant with direct steam generation in the receiver (single cycle). Charging is done by means of live steam and feed water and normal discharging by a series of flash evaporators [7.27]. In addition, at times of very high demand, the storage unit may act as sliding-pressure storage if insolation is low or zero or as feed water storage if insolation is medium to low.

An arrangement with internal steam generation is envisioned for the first USSR solar power station of 3 to 5 MW electrical power in the Crimean area [7.28].

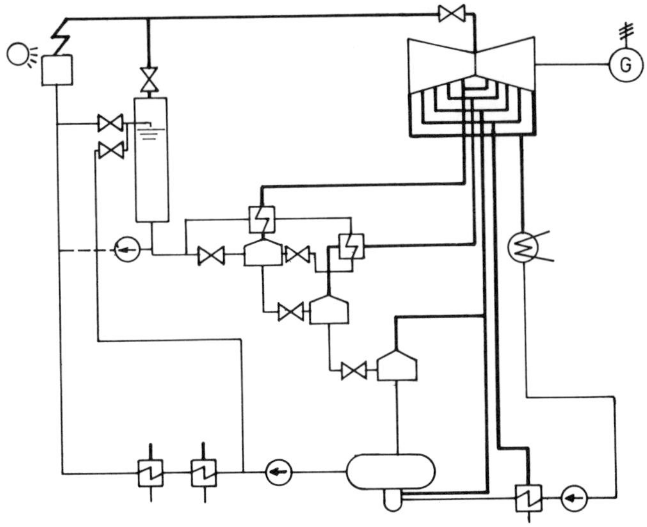

Fig. 7.27. Single cycle solar steam power plant with saturated steam storage [7.27]

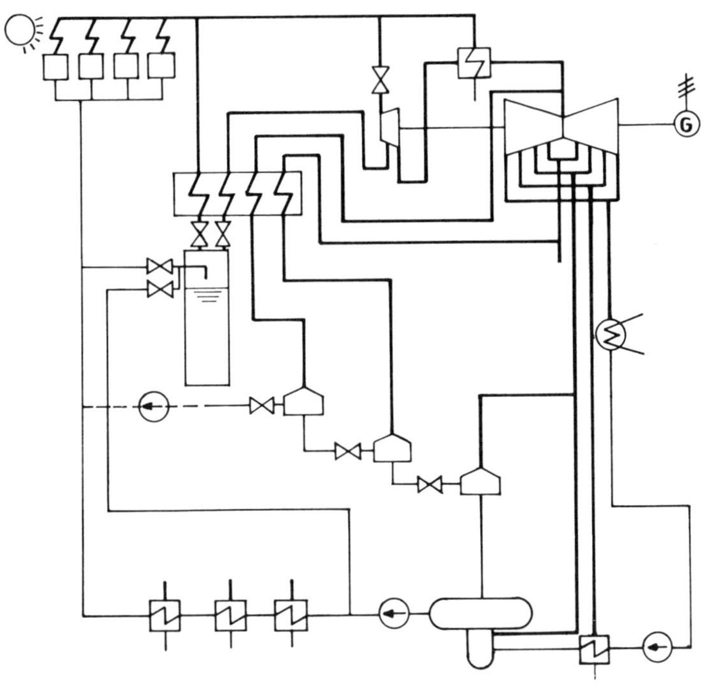

Fig. 7.28. Solar steam power station with superheated steam storage [7.27]

7.6.3. Single-cycle Solar Steam Plants with Indirect Thermal Storage

Fig. 7.28 shows a proposal that is similar to Fig. 7.27. This design involves direct steam storage but with an additional superheat storage system of the indirect type (for example, by means of heat transfer oil or molten salt) [7.27].

The indirect thermal storage of the 10 MW solar pilot plant at Barstow, California, is shown in Fig. 7.29. There is a single cylindrical tank of 3058 m^3 which contains 6100 tonnes of granite rock and 712 m^3 Caloria HT-43 heat transfer oil. The Thermocline storage tank is charged by means of live steam of 101 bar, 510°C which has been desuperheated to 348°C, the condensate leaving at 226°C. In the discharge mode, feedwater of 127°C enters the discharge heat exchanger and slightly superheated steam of 277°C, 27 bar is generated and piped to the dual-admission turbine [7.29].

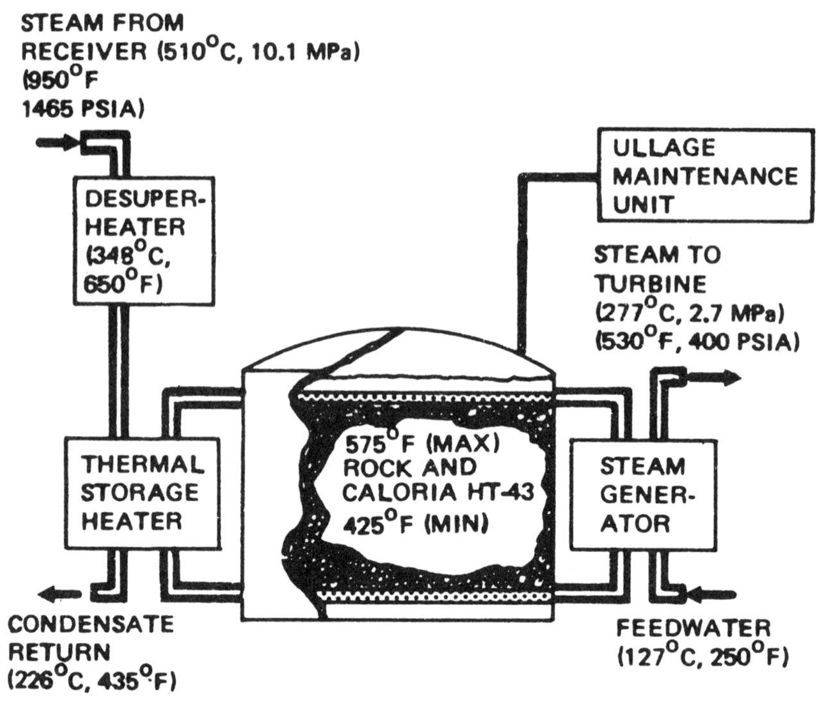

Fig. 7.29. Indirect thermal energy storage of the 10 MW solar plant at Barstow, California [7.29]

7.6.4. Double-cycle Solar Steam Plants

Fig. 7.30 presents the principal flow sheet of a double-cycle solar steam power plant with TES in the primary (heat transfer) cycle. The heat transfer

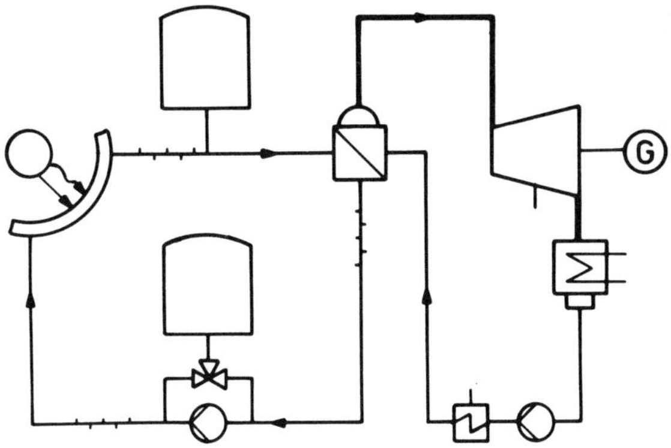

Fig. 7.30. Double cycle solar steam power plant with thermal energy storage in the primary heat transfer cycle [1.29]

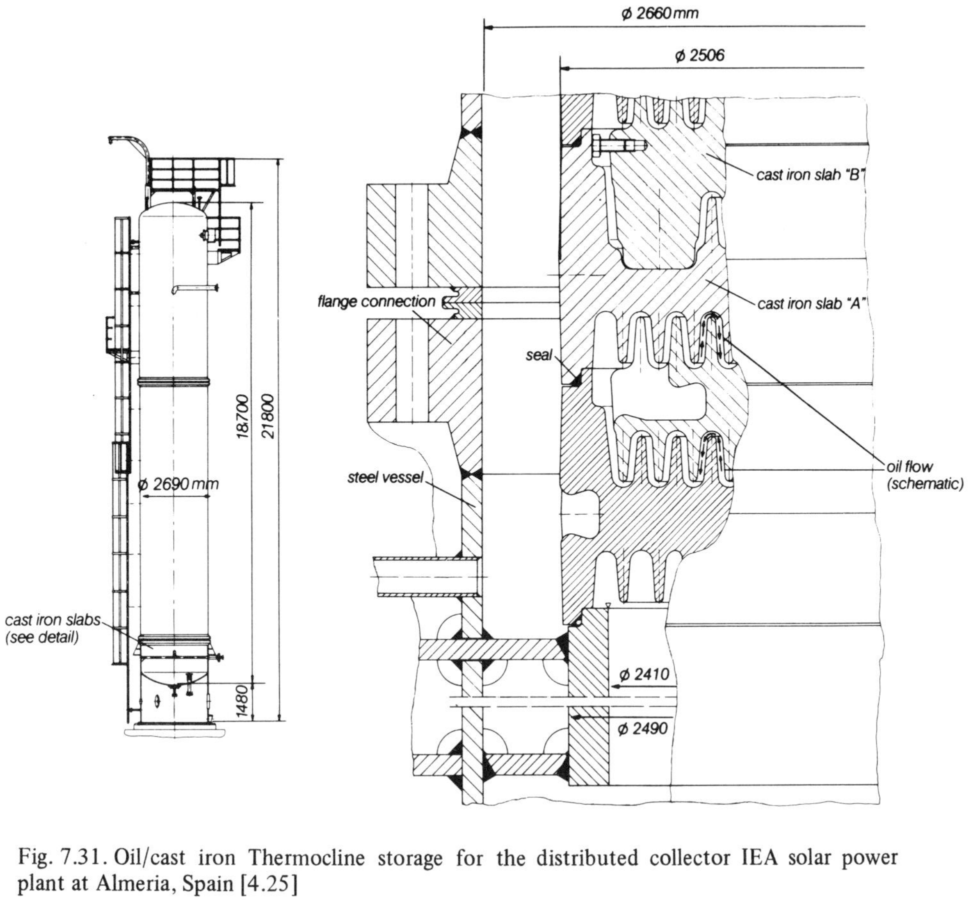

Fig. 7.31. Oil/cast iron Thermocline storage for the distributed collector IEA solar power plant at Almeria, Spain [4.25]

medium (heat transfer oil, molten salt, sodium) usually acts also as storage medium. There is a hot and a cold storage vessel [1.29] or a Thermocline arrangement.

Various heat transfer and storage media are in use:

- sodium in the 500 kW solar tower power plant of the International Energy Agency (IEA) at Almeria, Spain [7.30];
- heat transfer oil with labyrinth-shaped cast iron slabs (in order to increase the volumetric storage capacity and to minimize oil content—that is to be exchanged from time to time) in the Thermocline storage vessel (Fig. 7.31) of the 500 kW distributed collector solar power plant of the IEA in Almeria [4.25];
- HITEC in the French solar power plant THEMIS;
- water in small, low-temperature solar power plants which employ low-boiling liquids as working media. Fig. 7.32 shows the flow sheet of a 10 kW plant of this type [7.31].

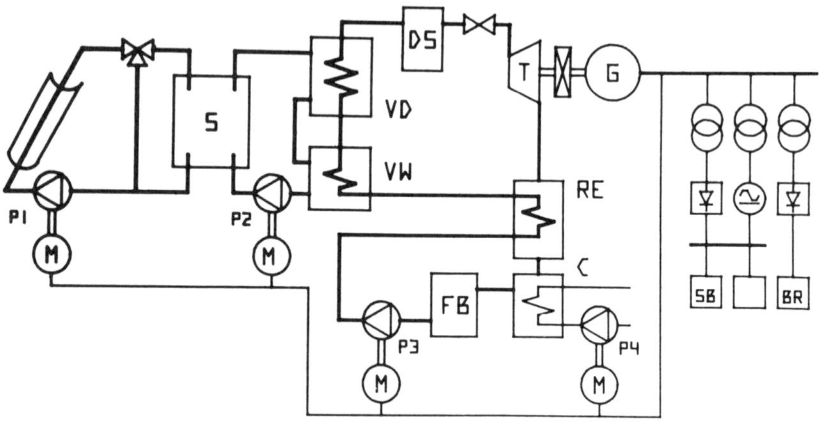

Fig. 7.32. Hot water storage of a 10 kW solar power plant [7.31]
RE recuperator, *S* storage tank, *VW* pre-heater, *VD* vaporizer, *DS* vapour storage, *T* turbine, *C* condenser, *FB* frigen reservoir, *P1, P2* collector pumps, *P3* frigen pump, *P4* coolant pump, *M* pump motors, *SB* auxiliary starting-up battery (110 V), *BR* control system battery (24 V), *G* generator

7.6.5. Solar Open-cycle Gas Turbine

Fig. 3.3 gives the basic flow sheet of a solar open Brayton cycle plant. High-temperature thermal storage by solids is used rather than pneumatic storage (as per Section 7.3.1), because the working pressure of the open-cycle gas turbine is much lower than the optimum design pressure of a pneumatic store [7.32].

Storage material may be rock, magnesia or cast iron in a Thermocline arrangement. Cast iron is more expensive, but it only requires a much smaller pressure vessel and may be economical in terms of overall cost. Charging is done by hot air from the receiver using a blower. During discharge, the storage unit heats up the working medium air (that is, it takes the function of the receiver).

7.6.6. Solar Closed-cycle Gas Turbine

Integrated TES for solar closed Brayton cycle plants (see Fig. 7.33) will use a high-temperature solid-body storage system similar to the one that is used in open-cycle plants. The working pressure will be much higher than those of an open-cycle system. Therefore, the overall cost advantage of the cast iron structure versus magnesia bricks will be even more pronounced.

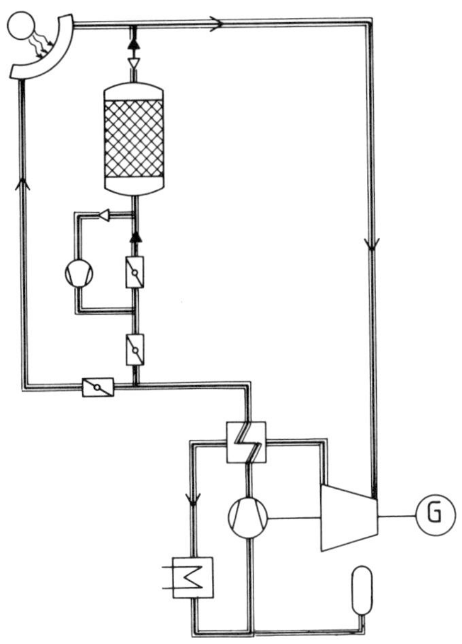

Fig. 7.33. TES in a solar closed-cycle gas turbine plant

Boeing Engineering and Construction has recently investigated TES for solar gas turbine power plants which use helium as the working medium [7.32]. One of the possibilities investigated was sensible heat storage of solids in an internally insulated pressure vessel with helium of 816/440°C and 37 bar (550 psi) as the heat transfer agent. A pressure vessel having a net volume (measured at the inside of the thermal insulation) of about 1200 m^3 is required for the plant which has a net electrical power output of 50 MW and a discharge duration of 6 hours. The full power output is thus extended from 5 p.m. to 11 p.m.

7.7. Conclusions

Although TES in power plants declined in importance over the last several decades, there are incentives for its future use and these incentives are briefly as follows:

- increased parallel use of inexpensive fuel (nuclear fuel, solar radiation) and expensive fuel (oil, gas);
- increased importance of constant load of nuclear reactors;
- increased utilization of power plants with high investment cost (nuclear and solar).

Although the basic principles of TES are rooted in the last century, there nevertheless will be a bright future for TES regardless of whether the energy of the future will be nuclear fission, nuclear fusion, solar radiation, or a mixture of these.

8. Thermal Energy Storage in Vehicles

8.1. General

Application of energy storage to vehicles is a special case of energy storage for power generation in that there is usually a local discrepancy between energy supply and demand in addition to the temporal discrepancy. Charging is performed in a charging station, while discharge occurs during travel. Additional charging may be done during braking of some energy storage vehicles, this is particularly so of vehicles with electric drive and is in analogy to wire-bound electric vehicles with feedback to the electric grid during braking.

However, there are energy storage vehicles without this local discrepancy. These are the so-called hybrid vehicles for which charging of the store is achieved by a thermal engine during travel and discharge power is used for purposes of overcoming peak loads (such as starting, accelerating, in making ascents) and/or for environmental reasons such as travel through towns and other congested areas.

The volume as well as the weight (mass) of the store is decisive for the application of energy storage to vehicles. The exergy density of thermal storage systems may be taken from Fig. 2.18. The dependence of mass-specific exergy on the upper working temperature or pressure is shown in Fig. 8.1. Not only the mass of the storage medium itself but that of the cylindrical pressure vessel also is accounted for. Wall thickness was assessed by the German Code using fine grain steel (WStE47 with a yield stress of 47 kgf/mm^2 = 461 MPa at 20°C and 294 MPa at 300°C; design stress = 1.1 p, $s \leqslant 70$ mm, $\nu = 1.5$, $\upsilon = 0.8$; $V_B^* = 2$; see Section 4.1; minimum vessel weight 100 kg/m^3).

Pressurized hot water storage and pressurized air storage show relatively low density but high mass-specific exergy. This is also true for some latent heat and thermochemical storage systems. Safety, high charge and discharge power and ease of charging are further criteria for selecting a storage system for vehicles.

Thermal (and compression) energy storage for vehicles is in competition with electric energy storage by means of electrochemical batteries (accumulators), with mechanical energy storage by means of fly-wheels and with fuel storage.

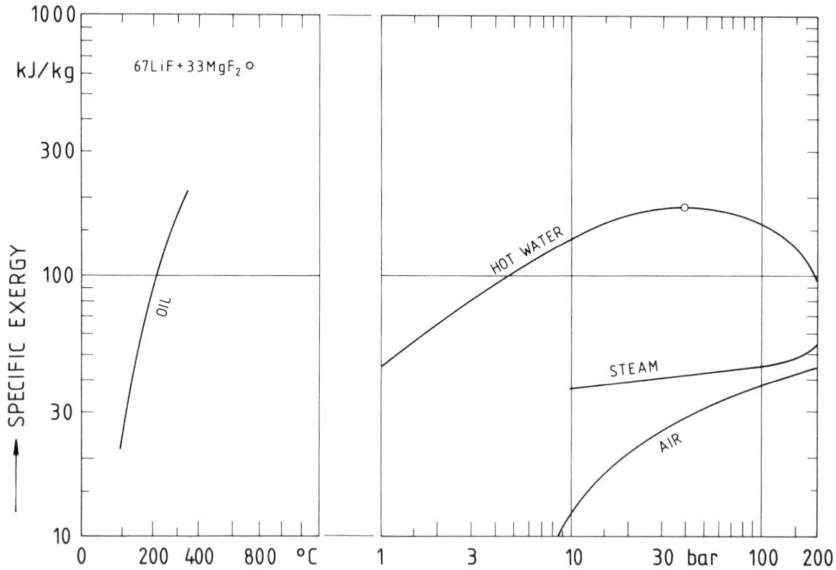

Fig. 8.1. Mass-specific exergy of storage media

8.2. Air and Sea Vehicles

Thermal energy storage for air and sea vehicles is limited to the special cases of extremely short travel time and/or distance and very high discharge power. Sliding-pressure storage is superior to electrical and mechanical storage in such cases.

8.2.1. Hot Water Rockets

The hot water rocket was used during World War II by the Germans as booster for starting the Me-262 interceptor [8.1]. It was later used as mail rocket and for meteorological purposes. The advantages of these rockets are inexpensive operation, possibility of recharging and long life.

The hot water rocket consists of a pressure vessel which is filled with pressurized hot water. This water is discharged into a jet according to the expansion principle. Part of the water evaporates in the jet. The specific impulse of these rockets is lower than that of chemical rockets. (For 50 bar storage pressure and 1 bar ambient pressure, the adiabatic enthalpy drop is 1155 – 1020 = 135 kJ/kg; the theoretical discharge velocity is $u = \sqrt{(2)(135\,000)} = 517$ m/s; and the specific impulse is 517 N per kg/s mass flow). However, the final velocity is sufficient for several applications.

8.2.2. Steam Catapults

Aircraft starting from air carriers are accelerated to starting velocity by catapults over a distance of about 100 meters in about 2.5 seconds. This requires an energy of about 100 MJ and, therefore, a power of 100/2.5 = 40 MW.

Sliding-pressure steam storage is used for this purpose, the discharge steam being applied to a piston via a quick-acting valve. The storage vessels are either 50% filled with hot water (wet storage), or a pure steam cushion storage without water is used (dry storage). Working pressures are, for instance, 67/50 bar (1000/750 psia), requiring 54 m^3 vessel volume for wet storage and 76 m^3 for dry storage (the steam consumption per launch is 862 kg) [8.2].

Since the number of cycles is rather high and the speed of discharge is high too, the need for a fatigue analysis including the thermal stresses is prominent.

8.2.3. Torpedoes

High pressure air storage has been applied to torpedoes since the middle of the 19th century (Whitehead torpedoes [8.3]). The air store supplies a pressurized air engine which acts via a gear box on two screw propellers which rotate in opposite senses. Typical data are: 5 m (length), 0.45 m (diameter) and 530 kg (starting weight). See Fig. 8.2.

Storage pressure is about 150 bar, being throttled to 38 bar before the pressurized air engine. The range is 6 km at a speed of 27 knots (50 km/h), or 1 km at 42 knots (78 km/h). Improvements of range and velocity have been achieved by heating up the air during travel (after throttling).

Fig. 8.2. Whitehead torpedo (Vienna Museum of Military History)

8.3. Rail Vehicles

The main reasons for applying thermal energy storage to locomotives instead of using the fired steam or diesel locomotives are a reduction in operation cost because of the low fuel cost of the stationary boiler and a reduction in maintenance cost and—in special cases—safety considerations.

8.3.1. Pressurized Air Locomotive

Pressurized air locomotives are used in mines when electric locomotives cannot be used because of danger of explosion. Their advantages are: no waste gas (rather there is a fresh air supply), no heat generation (rather there is a cooling effect by way of cold air exhaust and heat transfer) and simple operation and maintenance.

Fig. 8.3 shows the arrangement. Air taken from bottles with 200 bar storage pressure is throttled to the working pressure of 14 to 20 bar, expanded in the HP cylinder, reheated by ambient air in a tube bundle heat exchanger and further expanded in the LP cylinder. The exhaust air is used in an ejector for sucking ambient air through the reheater [8.4]. Besides this compound engine (with double-acting HP and LP cylinders), small locomotives are equipped with twin engines (double-acting, single-expansion) or with four cylinder engines (single-action, single-expansion). Charging takes place with flexible hoses at charging stations which are supplied with pressurized air from a compressor located above ground.

During discharge (in particular, during fast discharge) all components in contact with the pressurized air will cool down and may even become covered

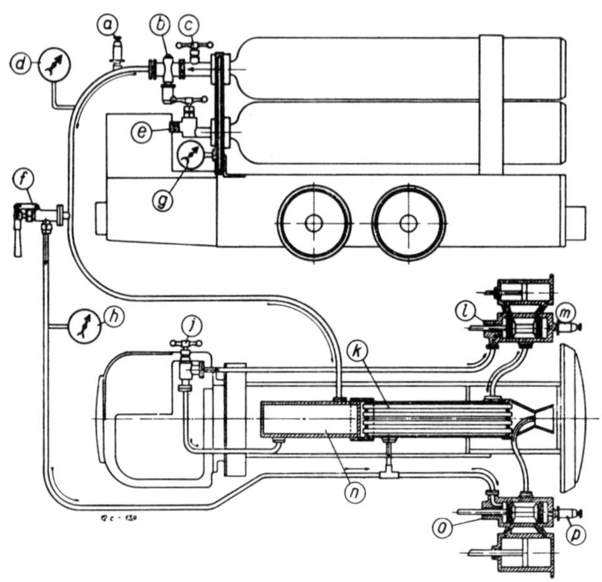

Fig. 8.3. Flow sheet of a pressurized air locomotive [8.4]
a safety valve behind pressure control valve, *b* pressure control valve (discharge valve), *c* shut-off valve, *d* pressure indicator (range 0–36 bar), *e* charge valve, *f* fresh air valve, *g* storage pressure indicator, *h* pressure indicator (range 0–16 bar), *j* governor valve, *k* reheater, *l* HP cylinder, *m* safety valve before HP cylinder, *n* intermediate vessel, *o* LP cylinder, *p* safety valve before LP cylinder

Fig. 8.4. 70 HP pressurized air locomotive [8.4]

by hoarfrost. This will reduce heat transfer from ambient air and will reduce capacity also. Storage pressure will rise slightly during stops. This rise stems from temperature equalization with ambient.

The pressurized air locomotives which have been built so far have storage pressures of up to 200 bar, storage volumes of up to 25 m^3 in one to three pressure vessels and ranges of up to about 6 km. Fig. 8.4 shows a locomotive of 70 HP (51 kW) and 12.5 kN driving force, with chain drive.

The measured specific output of a locomotive with 120 bar upper and 23 bar lower storage pressure was about 10 000 kJ/m^3 [8.5]. The theoretical output in this pressure range is about 42 000 kJ/m^3. Therefore, the discharge efficiency is about 24%. Losses consist of three main parts, the parts being of about equal size: Throttling losses at the pressure-reducing valve, losses which stem from non-isothermal expansion; and throttling losses, mechanical losses and exhaust losses of the engine. Thus, the efficiency of the pressurized air locomotive is lower than that of either the electric locomotive, or the electric storage (accumulator) locomotive, or the diesel locomotive.

8.3.2. Thermochemical Storage Locomotive

The thermochemical storage principle has been applied to street cars and locomotives (Chapter 1, Figs. 1.18 and 1.19) also. Fig. 2.16 D shows a flow sheet of the natron locomotive of Honigmann [8.6, 8.7] in principle. Exhaust steam reacts with the NaOH, reducing its concentration from 82% to about 65% and producing heat which in turn generates steam of 5.5 bar. This thermal storage system not only eliminates combustion on board but exhaust as well.

8.3.3. Steam Storage Locomotive

The steam storage locomotive has a sliding-pressure storage vessel in lieu of a boiler that is fired with coal or oil. Therefore, there is no smoke and no sparks. The steam storage locomotive may therefore be used by factories in which there is a risk of fire. It is also suitable for shunting duty where high peak power but low average power (only 7.5 to 10% of peak power) is required. Charging takes from 10 to 30 minutes and charging stations are supplied directly from a steam boiler, from a steam grid or from a charging store which is, in turn, charged directly of indirectly via a heat exchanger [8.8–8.14].

The conventional "fireless" locomotive has storage pressures that are equal to the maximum working pressure of about 13 bar to less than 3 bar. Effective exergy storage capacity is limited (being only about 3 to 2.2 kWh/m^3) and the discharge efficiency is rather low (11 to 15% exergetic and 3.5 to 4.8% energetic). Similarly, the range is rather small, being about 10 to 20 km (depending on load).

Based on the invention of Dr. Lamm, the "fireless" locomotive was first put into operation as street car in New Orleans in 1873. Already in 1875–1876, street cars, designed by L. Francq, using throttling and superheating of the working steam, were put on duty between St. Augustin and Neuilly in France and, somewhat later, in Indonesia, the former Batavia [8.10]. Superheating was again abandoned in the many "fireless" locomotives that were built in the following

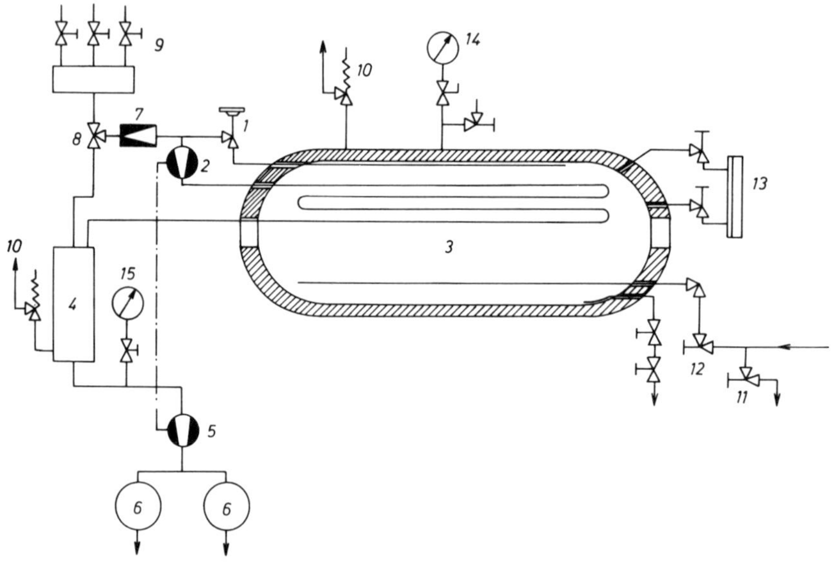

Fig. 8.5. Principle of the high-pressure steam storage locomotive (the Gilli locomotive) [8.12] *1* main shut-off valve, *2* pressure reduction valve, *3* high-pressure store, *4* low-pressure buffer store, *5* governor valve, *6* steam engine cylinder, *7* pressure reducing valve, *8* over-flow valve, *9* steam supply for auxiliaries, *10* safety valve, *11* drain, *12* charging valve, *13* water level gauge, *14* pressure indicator, store, *15* pressure indicator, buffer store

decades in Germany and other countries. They had pressures of up to about 30 bar, storage volumes of up to 23 m^3 and weights of up to 70 tonnes [8.10].

During the 1930s the Wiener Lokomotivfabrik Floridsdorf at Vienna introduced a high-pressure steam storage locomotive that had been developed by P. R. Gilli [8.8–8.14]. Fig. 8.5 shows the principle that is involved [8.10, 8.12]. Steam taken from the high pressure storage vessel (typically of 60 to 100 bar) is throttled to working pressure (about 14 to 17 bar) in a pressure reduction valve. It is then re-introduced into the storage drum, superheated in its steam part by condensing high pressure steam and finally lead via a governor valve to the engine. There is a small buffer store between superheater and regulator to soften working steam pressure variations–especially during start and low speed. Fig. 8.6 shows the process for full and half-loaded stores in an enthalpy-entropy diagram. Compared to the conventional "fireless" locomotive, the storage density is higher here. The discharge efficiency of the steam engine is about double that of the "fireless" locomotive because of the constant working pressure and the superheating of the steam. The range is about three times that of the "fireless".

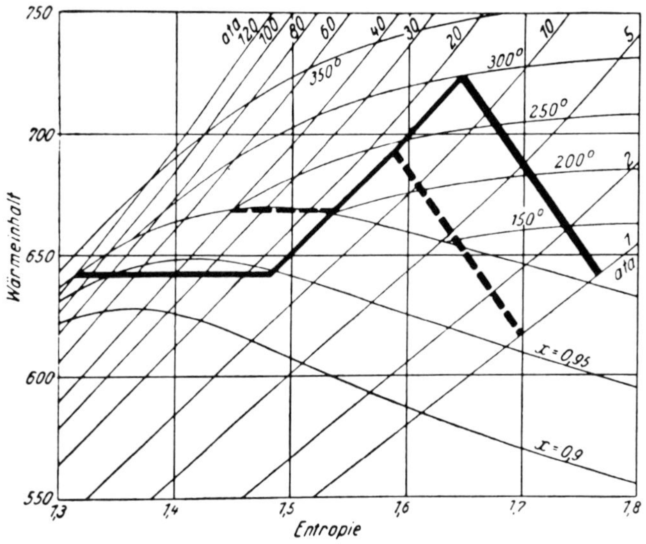

Fig. 8.6. Schematic expansion diagram of a high-pressure steam storage locomotive (enthalpy –kcal/kg–versus entropy–kcal/kg K) [8.9]

The first locomotive ot this type was built in 1934 for the gas works of the Vienna municipality. It was used to haul coal trains (of 1500 tonnes) from the main station over a distance of 1.5 km that included a ramp with a 17°/₀₀ grade which thus required a driving force of up to 200 MN. This was a ten-wheeler (D1) of 82 tonnes weight with a storage pressure of 118 bar [8.9]. Between 1934 and

1963 about 40 of these locomotives were built by the Wiener Lokomotivfabrik Floridsdorf which was owned from 1939 to 1946 by Henschel & Sohn of Kassel, Federal Republic of Germany, and since 1958 by the Simmering-Graz-Pauker AG (SGP) of Vienna. Most of these locomotives were six-wheelers, having storage pressures of between 40 and 100 bar; almost all of them had welded pressure vessels [8.13]. Fig. 8.7 shows one of these locomotives that was built in 1942 [8.14].

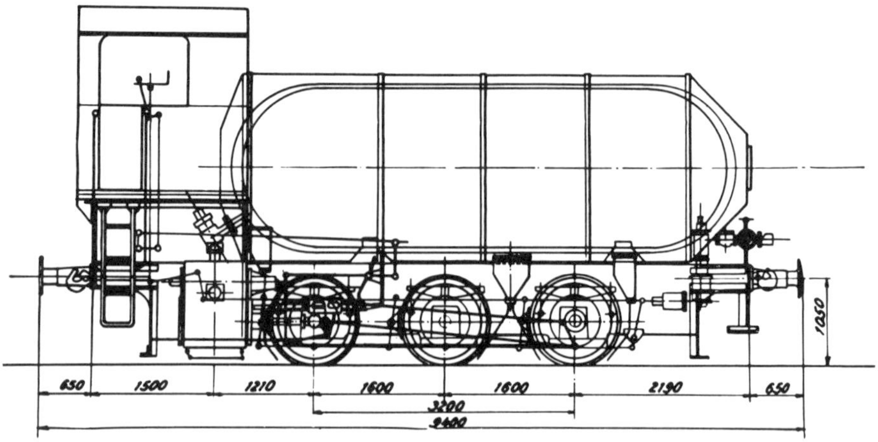

Fig. 8.7. High-pressure steam storage locomotive (54 tonnes, 86 bar, 1000 HP) [8.14]–dimensions in mm

Fig. 8.8. High-pressure steam storage locomotive, 85 bar, built in 1961/73 by SGP, Vienna (courtesy SGP)

From 1953 onwards, additional units were built by Henschel & Sohn at their Kassel factory; furthermore, four six-wheelers of 85 bar storage pressure (Fig. 8.8) were built between 1961 and 1973 by SGP for the Austrian refinery ÖMV at Schwechat near Vienna.

In light of the high costs of fuel and maintenance of diesel engines [8.10], high-pressure steam storage locomotives should, even today, be cost-effective (when compared to the diesel) for shunting duty in large factories and in railway stations without electric traction. Moreover, in the United States, the high-pressure steam storage locomotive has also been proposed for charging by solar energy [8.15].

8.4. Road Vehicles

The incentives for the application of secondary energy storage to road vehicles are oil substitution, reduction of environmental problems (especially in towns), coverage of peak loads for start-up and accelerating, and recovery of brake energy. Providing the necessary infrastructure for recharging may be problematic in the initial phase for most of the systems.

The exhaust from steam storage may be a disadvantage for road vehicles because of possible fogging and icing problems. However, pressurized air storage, thermochemical storage and latent heat storage in combination with the Stirling engine have been proposed and some prototypes have been built.

8.4.1. Road Vehicles with Pressurized Air Storage

Pressurized air storage for road vehicles has recently been proposed. A small prototype with a rotary engine of 4.5 bar working pressure and single expansion was built in Austria. Brake energy is recovered via a compressor which is also used for recharging by means of an electric motor. Alternatively, the pressurized air bottles may be exchanged, or they may be recharged by a stationary compressor. Since electric energy is normally used for recharging, the pressurized air car may be regarded as one particular embodiment of the electric car [8.16].

Air or nitrogen storage has also been proposed for recovering brake energy as well as for covering peak load for starting and accelerating [8.17]. In this way, the engine power needed for the same acceleration may be reduced by a factor of 2 or more. The necessary storage capacity is rather small; for starting and accelerating a large passenger car of 1800 kg to a speed of 60 km/h only 0.1 kWh (360 kJ) of mechanical energy are required. Fig. 8.9 shows the principle. The internal combustion engine acts on a hydraulic oil pump with variable displacement which also fulfills the function of the gear box. The pressurized oil drives an oil motor which acts directly on the wheels; the oil motor operates as pump during braking. The HP and LP sides of the oil motor/pump are connected to the "hydro store" having 350 bar maximum pressure. As a consequence of the constant engine load and the continuous transmission, the average efficiency of the engine is better than that of conventional drives. In addition, 50% of brake energy can be recovered.

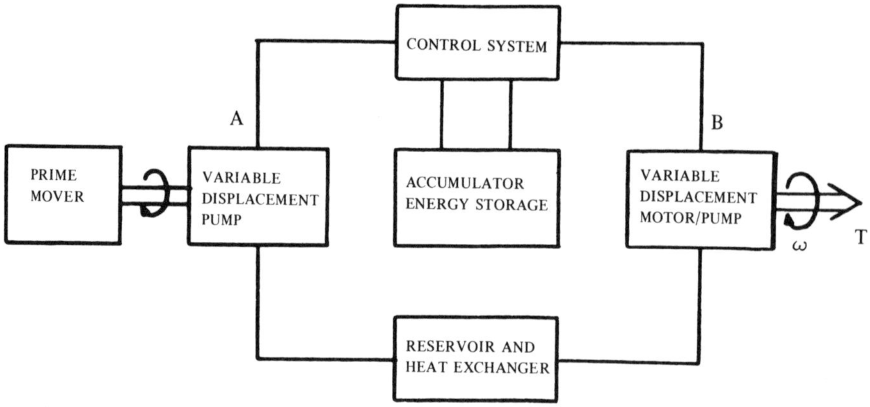

Fig. 8.9. Hybrid drive of a road vehicle with pressurized gas storage and hydraulic transmission [8.17]

8.4.2. Road Vehicles with Thermochemical Storage

Use of thermochemical storage for road vehicles has recently been proposed as "Concentration Difference Energy" (CDE) system. A small vehicle with a steam engine of 0.4 kW net power (driven with steam of initially 5 bar, finally 2.5 bar) has been built and tested in Japan [8.18].

Contrary to the natron locomotive described in Section 8.3.1, an aqueous solution of 40% $CaCl_2$/60% LiCl is used. A further difference is the possibility of recharging not only by thermal energy but alternatively by mechanical or electrical energy via the steam engine acting as a steam compressor. The vehicle may thus be regarded as an electric vehicle.

8.4.3. Road Vehicles with Storage of Latent and Sensible Heat

The use of high-temperature latent heat storage for Stirling engines has been proposed in Europe for buses [8.19] as well as for passenger cars [8.20]. Similar proposals have been made in the United States [8.21, 8.22]. The conceptual design of Sigma Research Inc. is based on a car with a 160 km (100 miles) range at a speed of 88 km/h (55 mph) or of 120 km (75 miles) range in urban traffic. Assuming a 40% efficiency for the Stirling engine, this requires a thermal storage capacity of 102 kWh. Fig. 8.10 shows the principle. An eutectic salt mixture (LiF + MgF_2) with a melting point of 741°C is used in the temperature range of between 825°C and 600°C (that is, including some of the sensible heat of solid and molten salt). In this way, the storage capacity is increased by one third above that of a pure latent heat store. But a higher design temperature and a decreasing working temperature also has to be accepted. The salt is contained in tubes of 316 S stainless steel that are arranged in a horizontal cylindrical storage vessel made of Inconel 617. From there, heat is transferred by a two-

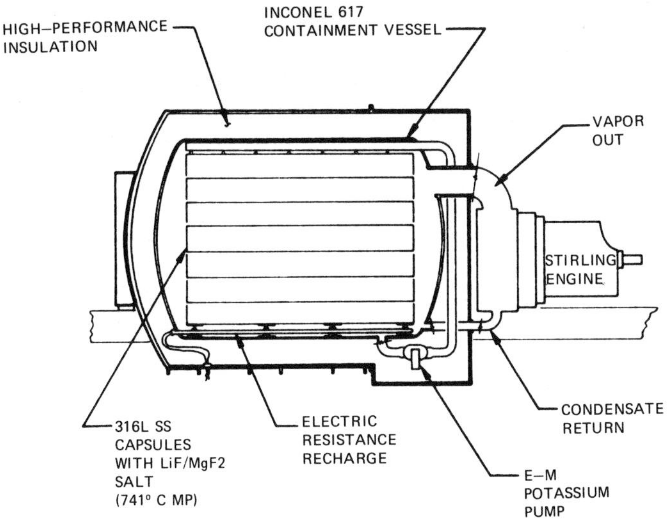

Fig. 8.10. Drive of a road vehicle with latent TES and a Stirling engine [8.21]

phase potassium loop (consisting of an electro-magnetic potassium recirculation pump, a potassium steam line and a condensate return line) to the heater of the Stirling engine. The salt store has a mass of 500 kg and a volume of 0.4 m^3. It is thermally insulated with evacuated, highly porous silica [8.21].

Other designs are based on the heat pipe principle [8.19, 8.20, 8.22]. Variation of pressure of the heat pipe fluid is used for load control of the engine. If a swashplate engine is used, then load control will be achieved by varying the displacement. Fig. 8.11 [8.20] shows a proposed arrangement of the heat storage in a passenger car.

The charging of the store is done by a stationary gas burner, or preferably by electric energy. Thus, this vehicle may also be regarded an electric car. The

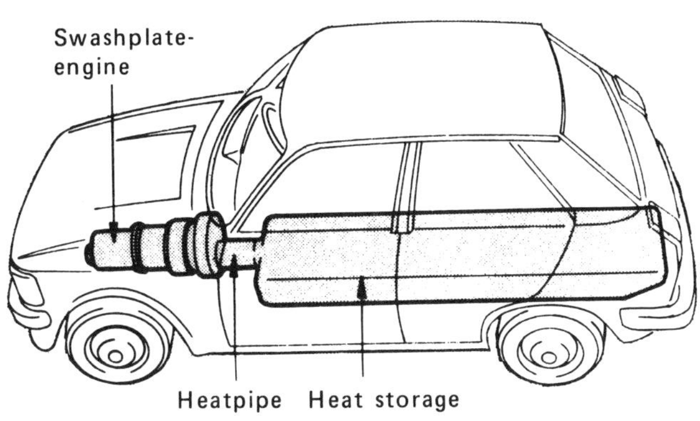

Fig. 8.11. General arrangement of a TES passenger car [8.20]

overall efficiency of the car is given as 20% for gas burner charge and as 32% for electric charging.

Aside this rather involved proposal with latent heat storage, pure sensible heat storage systems have also been proposed and built. Their ranges are shorter (in accordance with the lower exergy density of sensible heat systems). In one car having a steam engine [8.23], oil of between 315 and 177°C (600 and 350°F) is the storage medium. The range is 56 km (35 miles) at 88 km/h (55 mph). The charging station uses a gas burner. The use of solar energy has been proposed in this connection.

In principle, the above thermal storage cars may also be operated as hybrids – for instance by means of a small furnace on board which operates on relatively inexpensive fuel (such as fuel oil) and recharges a relatively small store (of only one tenth of the capacity of a pure storage vehicle).

8.5. Conclusions

In transportation TES has been applied to air, sea, rail and road vehicles. The main emphasis of future energy technologies for transportation will be on oil substitution. It will be achieved by fuel switching, electrification, or energy storage. The main application of TES to transportation in the future should be to road vehicles and (partly) to locomotives in unelectrified railway systems.

References

Chapter 1

1.1. ETE-MEE-EGI Conference on Peak-Load Coverage (Budapest, Hungary, November 18–20, 1969).

1.2. Speichersysteme für Sekundärenergie (Stuttgart, 1974). (VDI-Berichte 223.) Düsseldorf: VDI-Verlag. 1974.

1.3. Proceedings of the Second Annual Thermal Energy Storage Contractor's Information Exchange Meeting (September 29–30, 1977, Gatlinburg), CONF-770955. Springfield, Va.: NTIS. 1977.

1.4. Proceedings of the Third Annual Thermal Energy Storage Contractor's Information Exchange Meeting (December 5–6, 1978, Springfield, Va.), CONF-781231. Springfield, Va.: NTIS. 1978.

1.5. Thermal Energy Storage, Report of a NATO Science Committee Conference (Turnberry, Scotland, March 1–5, 1976) (Kovach, E. G., ed.). Oxford: Pergamon Press. 1976.

1.6. Rationelle Energienutzung durch Wärme-Speicherung (Stuttgart, 1977). (VDI-Berichte 288.) Düsseldorf: VDI-Verlag. 1977.

1.7. Thermal Storage and Heat Transfer in Solar Energy Systems (ASME Winter Annual Meeting, San Francisco, 1978) (Kreith, F., et al., eds.). New York: ASME. 1978.

1.8. Energy Storage–A Vital Element in Mankind's Quest for Survival and Progress (Proceedings of the First International Assembly on Energy Storage, Dubrovnik, May 27–June 1, 1979) (Silverman, J., ed.). Oxford: Pergamon Press. 1980.

1.9. Peaking Power Generation (ASME Winter Annual Meeting, Chicago, 1980) (Miliaras, E. S., ed.). New York: ASME. 1980.

1.10. International Seminar on Thermochemical Energy Storage (Stockholm, January 7–9, 1980) (Wettermark, G., ed.). Stockholm: Swedish Council for Building Research. 1980.

1.11. Thermal Storage of Solar Energy (Proc. Int. TNO Symposium, Amsterdam, November 5–6, 1980) (Den Ouden, C., ed.). The Hague-Boston-London: M. Nijhoff. 1981.

1.12. Proceedings of the DOE Thermal and Chemical Storage Annual Contractor's Review Meeting (October 14–16, 1980, McLean, Va.), CONF-801055. Springfield, Va.: NTIS. 1981.

1.13. Energy Storage and Transportation–Prospects for New Technologies (Lectures of a Course Held at the Joint Research Centre, Ispra, October 22–26, 1979) (Beghi, G., ed.). Dordrecht-Boston-London: D. Reidel. 1981.

1.14. Proceedings of the International Conference on Energy Storage (Brighton, 1981), 2 Vols. (Stephens, H. S., Jarvis, B., eds.). Cranfield-Bedford: BHRA Fluid Engineering. 1981.

1.15. Thermal Energy Storage (Lectures of a Course Held at the Joint Research Centre, Ispra, June 1–5, 1981) (Beghi, G., ed.). Dordrecht-Boston-London: D. Reidel. 1982.

1.16. Second International BHRA Fluid Engineering Conference on Energy Storage for Energy Management (Stratford-upon-Avon, 1983). Cranfield-Bedford: BHRA Fluid Engineering. 1983.

1.17. Pauer, W.: Energiespeicherung. Dresden-Leipzig: Th. Steinkopf. 1928.

1.18. Musil, L.: Die Wirtschaftlichkeit der Energiespeicherung für Elektrizitätswerke. Berlin: Springer. 1930.
1.19. Goldstern, W.: Dampfspeicheranlagen – Bau, Berechnung und Betrieb industrieller Wärmespeicher, 2. Aufl. Berlin-Göttingen-Heidelberg: Springer. 1963.
1.20. Goldstern, W.: Steam Storage Installations (International Series of Monographs in Mech. Engineering, Vol. 4). Oxford: Pergamon Press. 1970.
1.21. Criteria for Energy Storage R & D. Washington: National Academy of Sciences. 1976.
1.22. Wettermark, G., et al.: Storage of Heat–A Survey of Effects and Possibilities, Document D2: 1979. Stockholm: Swedish Council for Building Research. 1980.
1.23. Bylin, F.: Low Temperature Thermal Energy Storage: A State-of-the-Art Survey, SERI/RR-54-164. Springfield, Va.: NTIS. 1979.
1.24. Atterkvist, St.: Swedish Energy Storage Prospects 1979; D19: 1980. Stockholm: Swedish Council for Building Research. 1980.
1.25. Hall, E. W., et al.: Conceptual Design of Thermal Energy Storage Systems for Near Term Electric Utility Applications, Final Report, DOE/NASA/0012-79/2. Springfield, Va.: NTIS. 1979.
1.26. Steiner, D., et al.: Studie über thermische Energiespeicher für den Temperaturbereich 200 °C bis 500 °C, BMFT-FB-T 82-105. Karlsruhe: FIZ. 1982.
1.27. Euler, K.-J.: Energiespeicher. (Haus der Technik – Vortragsveröffentlichungen, Heft 451.) Essen: Vulkan-Verlag Classen. 1982.
1.28. Sauer, E., Zeise, R.: Energietransport, -speicherung und -verteilung. (Handbuchreihe Energie, Vol. 11.) Gräfelfing/Köln: Resch/TÜV Rheinland. 1982.
1.29. Gilli, P. V., Beckmann, G.: Thermal Energy Storage for Peaking Power Generation, [1.15], 265–300.
1.30. Proceedings of the DOE Physical and Chemical Energy Storage Annual Contractor's Review Meeting (August 23–26, 1982, Arlington, Va.), CONF-820827. Springfield, Va.: NTIS. 1982.
1.31. Berman, I. M.: Developments in Energy Storage Technology. Power Engineering, July 1983, 46–54.
1.32. Proceedings of the International Conference on Subsurface Heat Storage in Theory and Practice (Stockholm, June 6–8, 1983). Stockholm: Swedish Council for Building Research. 1983.

Chapter 2

2.1. Baehr, H. D.: Thermodynamik, 4th ed. Berlin-Heidelberg-New York: Springer. 1978.
2.2. Brown, G. B., Sliepcevich, C.: Practical Thermodynamics. Chemical Engineering Progress *48*, 493–496 (1952).
2.3. Föhl, C.: Über die Ladung von Ruthsspeichern. Ingenieur-Archiv *1*, 403–443 (1930).
2.4. Gilli, P. R.: Hochdruckdampfspeicher für feuerlose Lokomotiven. Dissertation, Technische Hochschule Wien, 1933.
2.5. Beckmann, G.: Dampfspeicher zur wirtschaftlichen Spitzenlastdeckung mit Kernkraftwerken. Dissertation, Technische Hochschule Graz, 1973.
2.6. Beckmann, G.: Die Entladekapazität von Gefällespeichern. Wärme *82*, 110–117 (1976).
2.7a. Hardegen, H.: Zur Berechnung der Arbeitskreise von Heißwasserspeichern. (DLR-Forschungsbericht 72-24.) Porz-Wahn: Deutsche Forschungs- und Versuchsanstalt für Luft- und Raumfahrt e. V. 1972.
2.7b. Hardegen, H.: Näherungsweise Berechnung der Zustände in Wasserspeichern mit siedendem und nichtkonstantem Inhalt bei Dampf- bzw. Dampf-Luft-Entnahme. Forsch. Ing.-Wes. *40*, 47–59 (1974).
2.8. Schmidt, E., Grigull, U.: Properties of Water and Steam in SI Units, 2nd Printing. Berlin-Heidelberg-New York-München: Springer/R. Oldenbourg. 1979.
2.9. Gilli, P. V., Beckmann, G.: Covering Peak Load by Means of Thermal Energy. (VDI-Berichte 236: Lines of Development in Energy Engineering/Energy Engineering Convention 1975.) Düsseldorf: VDI-Verlag. 1975.

2.10. Eisenmann, G., Hahne, E.: Energiespeicherung. Brennstoff-Wärme-Kraft *29*, 151–154 (1977).

2.11. VDI-Wärmeatlas, 2nd ed. Düsseldorf: VDI-Verlag. 1977.

2.12. Bauer, H. C., Alefeld, G.: Erhöhung der Wärmekapazität von Heißwasserspeichern. Wärme *83*, 35–36 (1977).

2.13. Ullmanns Encyklopädie der Technischen Chemie, 4th ed., Vol. 2: Verfahrenstechnik I (Grundoperationen). Weinheim/Bergstr.: Verlag Chemie. 1975.

2.14. Mancini, N. A., Simone, F.: Sensible Heat Storage, [1.15], 21–32.

2.15. Wood, R. J.: Thermal Energy Storage for the Recovery of Industrial Waste Heat, [1.15], 219–263.

2.16. Dubbel – Taschenbuch für den Maschinenbau, 15. Aufl. Berlin-Heidelberg-New York: Springer. 1983.

2.17. Kesselring, P.: Zur Energiedichte im Latentwärmespeicher – einige grundsätzliche Überlegungen, [1.6], 87–95.

2.18. Maru, H. C., et al.: Molten Salt Thermal Energy Storage Systems: Salt Selection, US/ERDA-Report C00-2888-1. Springfield, Va.: NTIS. 1976.

2.19. Petris, R. J., Claar, T. D.: High Temperature Molten Salt Thermal Energy Storage Systems for Solar Applications, [1.12], 60–64.

2.20. Schröder, J.: Energiespeicherung in Form von Wärme. Wärme *79*, 80–81 (1973).

2.21. Birchenall, C. E.: Heat Storage in Alloy Transformations, [1.12], 73–78.

2.22. Biswas, D. R.: Thermal Energy Storage Using Sodium Sulfate Decahydrate and Water. Solar Energy *19*, 99–100 (1977).

2.23. Nemecek, J. J., et al.: Demand Sensitive Energy Storage in Molten Salts. Solar Energy *20*, 213–217 (1978).

2.24. Pierce, B. L., et al.: Thermal Energy Storage. Proceedings of the 12th Intersociety Energy Conversion Engineering Conference (1977); Paper 779191; 1189–1194.

2.25. Schwind, H., et al.: Über die Verwendung von quergerippten Rohren in Latentwärmespeichern. vt (Verfahrenstechnik) *12*, 729–734 (1978).

2.26. Alefeld, G.: Energiespeicherung durch Heterogen-Verdampfung, Part I: Physikalisch-technische Grundlagen. Wärme *81*, 89–93 (1975); Part II: Ein Verfahren zur Fernwärmespeicherung und Spitzenstromerzeugung. Energie *27*, 180–183 (1975).

2.27. Niebergall, W.: Arbeitsstoffpaare (Arbeitsmittel und Absorptionsmittel) für Absorptions-Kälteanlagen und Absorptions-Kühlschränke. Mühlhausen/Thür.: Verlag Markewitz. 1949.

2.28. D'Ans-Lax: Taschenbuch für Chemiker und Physiker, 3rd ed., Vol. I: Makroskopische physikalisch-chemische Eigenschaften. Berlin-Heidelberg-New York: Springer. 1967.

2.29. Bauerle, G., et al.: Storage of Thermal Energy in Inorganic Oxides/Hydroxides, [1.10], 205–246.

2.30. Gintz, J. R., Pepper, J.: Technical and Economical Assessment of Phase Change and Thermochemical Advanced Thermal Energy Storage (TES) Systems. EPRI-Report EM-256 (1976).

2.31. Schmidt, E. W., Lowe, P. A.: Thermochemical Energy Storage Systems. Proceedings of the 11th Intersociety Energy Conversion Engineering Conference (1976); Paper 769116; 665–672.

2.32. Harth, R., et al.: EVA-ADAM System: A Method of Energy Transportation by Reversible Chemical Reactions, [1.15], 358–374.

2.33. Buchner, H.: Energiespeicherung in Metallhydriden. (Innovative Energietechnik.) Wien-New York: Springer. 1982.

Chapter 3

3.1. Oxley, A. J.: Methods and Means for Storing Energy. US-Patent 4.227.374, filed October 20, 1978.
3.2. Gilli, P. V., Beckmann, G., Schilling, F. E.: Nuclear Peak Load Plant with PCIV Thermal Energy Storage. Technical Meeting C 1.5, "Nuclex 78" (Basel, October 3–7, 1978).
3.3. Lehmann, J.: Klar zum Schwarzstart: Luftspeicher-Gasturbinen-Kraftwerke. Energie *34*, 51–52 (1982).
3.4. Marguerre, F.: Das thermodynamische Speicherverfahren von Marguerre. Escher-Wyss-Mitteilungen *6*, 67–76 (1933).
3.5. Pacault, P., et al.: Reducing the Cost of Energy of a PWR by the Addition of Thermal Storage and a Steam Pump. A/CONF. 49/P 582. 4th UN International Conference on the Peaceful Uses of Atomic Energy. Vol. 4, 165–176. Wien: IAEA. 1971.
3.6. Schröder, R.: Große Dampfkraftwerke. Vol. 2: Die Lehre vom Kraftwerksbau. Berlin-Göttingen-Heidelberg: Springer. 1962.

Chapter 4

4.1. Klapp, E.: Apparate- und Anlagentechnik. Berlin-Heidelberg-New York: Springer. 1980.
4.2. McLeish, R. D.: Pressure Vessel Cost Variation Depends on Shapes and Sizes. Process Engineering *1972*, 65–68.
4.3. Geipel, W., Zeitvogel, H.-R.: Die Zahlen sprechen für den Stahlzylinder. Energie *33*, 156–161 (1981).
4.4. Special Report: Worldwide Pressure Vessel Codes. Hydrocarbon Processing *57*, 89–101 (1978).
4.5. Beckmann, G.: Gravity-Type Steam Accumulator. US-Patent 4,027.690, filed Aug. 1975.
4.6. Västerås Stads Wärmekraftwerk AG: Wärmespeicherung – Technik und Wirtschaftlichkeit – Konventionelle Speicherung. Fernwärme International (FWI) *8*, 20–23 (1979).
4.7. Grigull, U.: Die Grundgesetze der Wärmeübertragung, 3. Aufl. Berlin-Göttingen-Heidelberg: Springer. 1963.
4.8. Dooley, J. L., et al.: A Feasibility Study of Underground Energy Storage Using High Pressure, High-Temperature Water, Prepared for the U.S. ERDA Under Contract No. E(04-3)-1243, January 1977.
4.9. Weber, O.: Das luftbetriebene Gasturbinenkraftwerk Huntorf. Brown Boveri Mitteilungen *62*, 332–337 (1975).
4.10. Margen, P. H.: Thermal Energy Storage in Rock Chambers–A Complement to Nuclear Power. 4th UN International Conference on the Peaceful Uses of Atomic Energy (Geneva, 1971); Vol. 4, 177–198. Wien: IAEA. 1972.
4.11. Barnstaple, A. G., et al.: Underground Thermal Energy Storage, Report No. 77113. Toronto: Ontario Hydro, Energy Environmental Studies Department. 1977.
4.12. Powell, J. D., Powell, J. R.: A System for Undersea Energy Storage. 15th Intersociety Energy Conversion Engineering Conference (1980); Paper 809047, 233–238.
4.13. Bauer, F.: Spannbetonbauten. Wien-New York: Springer. 1971.
4.14. ASME Boiler and Pressure Vessel Code, Section III, Div. 2: Code for Concrete Reactor Vessels and Containments. American Society of Mechanical Engineers, ACI Standard 359–374 (1981).
4.15. British Standard Specification for Prestressed Concrete Pressure Vessels for Nuclear Reactors. London: British Standard Institution BS 4973. 1973.
4.16. Caissons des réacteurs nucléaires en béton précontraint par armatures métalliques. Edité du Ministère du Développement Industriel et Scientifique (June, 1970); et: Circulaire du 15 Juin 1970 relative à l'application de la réglementation des appareils à pression aux caissons, des réacteurs nucléaires en beton précontraint. Edité du Ministère du Developpement Industriel et Scientifique.
4.17. Margen, P. H., et al.: Nordic Studies on Prestressed Concrete Pressure Vessels for Water

Reactors. 4th UN International Conference on the Peaceful Uses of Atomic Energy (Geneva, 1971), Vol. 2, 393–414. Wien: IAEA. 1972.
4.18. Bremer, F.: Entwicklung eines Spannbetonbehälters für Leichtwasserreaktoren und zugehörige Komponenten, Technische Mitteilungen, Krupp-Forschungsberichte, Heft 3; 71–78 (1973).
4.19. Fritz, K., Nemet, J.: A PCPV with Elastic Hot Liner. BNES International Conference on Gas-Cooled Reactors Today (Bristol, September 1982); Paper 26, Vol. 4, 39–48.
4.20. Beine, B., Gross, H., Schilling, F. E.: The Prestressed Cast Iron Vessel (PCIV)–Its Applicability for Gas- and Water-Cooled Nuclear Power Reactors and for Burst Protection. Nuclear Engineering and Design *28*, 387–399 (1974).
4.21. Beine, B.: Research and Development Program for Prestressed Cast Iron Pressure Vessels (PCIV) for High-temperature Reactors, 4th International Conference on Structural Mechanics in Reactor Technology (San Francisco, 1977), Paper H 1/7.
4.22. Gülicher, L., Schilling, F. E.: Further Fields of Application for Prestressed Cast Iron Pressure Vessels (PCIV). 4th International Conference on Structural Mechanics in Reactor Technology (San Francisco, 1977), Paper H 1/11.
4.23. Gilli, P. V., Beckmann, G., Schilling, F. E.: Thermal Energy Storage Using Prestressed Cast Iron Vessels (PCIV), ERDA-Report C00-2886-1. Springfield, Va.: NTIS. 1977.
4.24. Brandes, H., Gülicher, L.: Einsatz von Heißwasserspeichern in Kraftwerken zur Deckung der elektrischen Spitzenlast und des Fernwärmebedarfes, [1.6], 145–151.
4.25. 1981 PCIV: The First Prestressed Cast Iron Vessel (Company Brochure). Krefeld: Siempelkamp + Steinmüller GmbH. 1981.
4.26. Tamplyn, R. T.: Thermal Storage: Resisting Temperature Blending. ASHRAE J. *1980*, 65–70.
4.27. Scholz, F.: Seasonal Storage of Low Temperature Heat in Big Water Reservoir, [1.11], 221–232.
4.28. Uppsala Industriewerk, Uppsala Kraftwärme AB, Fyriskraft AB, Company brochure (1982).
4.29. Öfverholm, E.: The Swedish Seasonal Thermal Storage Programme with Application to the Building Sector, [1.11], 91–98.
4.30. Schöll, G.: Warmwasser-Großwärmespeicher, [1.2], 33–35.
4.31. Systems for Heat Storage. Studsvik Technical News 4 (1975).
4.32. Bounin, D., Beine, B.: Prestressed Cast Iron Pressure Vessels. Developments in Pressure Vessel Technology–4: Design for Specific Applications (Nichols, R. W., ed.). London-New York: Applied Science Publishers. 1983.

Chapter 5

5.1. Beckmann, G., Gilli, P. V.: Dampf- und Wärmespeicherung in der Industrie. X. Arbeitstagung Verfahrenstechnik (Graz, Austria; 1981), 187–195. Wien: Österr. Arbeitsgruppe für Apparatewesen und Verfahrenstechnik. 1981.
5.2. Goldstern, W.: Thermal Energy Storage in Industry and Power Stations, [1.14], 113–121.
5.3. Heiligenstaedt, H.: Wärmetechnische Rechnungen für Industrieöfen, 4. Aufl. Düsseldorf: Verlag Stahleisen. 1966.
5.4. Hausen, H.: Wärmeübertragung im Gegenstrom, Gleichstrom und Kreuzstrom, 2nd ed. Berlin-Heidelberg-New York: Springer. 1976.
5.5. Scheiblechner, G.: Veitscher Chimney-type Checker Block Packing Method for Glass Tank Regenerators. Glass Technology *23*, 223–229 (1982).
5.6. Dürselen, H.: Bemessung neuzeitlicher Koksofenregeneratoren. Techn. Mitteilungen *73*, 805–809 (1980).
5.7. Krupp-Koppers GmbH: Hot Blast Stoves. (Company Brochure.) Essen: 1980.
5.8. Katter, L. B., Rohnsen, P. J.: Applications of TES to Process Heat and Waste Heat Recovery in the Iron and Steel Industry, [1.4], 327–331.
5.9. Taylor, P. B.: Energy Audit Series No. 16: The Iron and Steel Industry. London: Department of Energy/Department of Industry. 1982.

5.10. Karas, P., et al.: Dry Coke Cooling–The Waagner-Biro Process. Iron and Steel Engineer *1981*, 49–54.
5.11. ENCOKE–A Process of Coke Dry Quenching from Krupp-Koppers GmbH and Waagner-Biro AG, Waagner-Biro AG Information; Wien, March 1982.
5.12. The Year-Book 1976 of the Coke Oven Manager's Association (C.O.M.A.). Mexborough, Yorkshire: Year-Book Ltd. 1976.
5.13. Rasworschegg, H.: Konstruktion und Betriebserfahrungen von Abgas-Kühlanlagen in Blas-Stahlwerken. Mitteilungen der VGB, Heft 74, Oktober 1961.
5.14. Koichi Horio, et al.: Waste Energy Recovery at Kashima Steel Works. Iron and Steel Engineer *1982*, 30–36.
5.15. Younger, R. N., Haworth, D. J.: Energy Conservation in the Steelplant. (Paper No. 7, Energy Conscious Iron- and Steelmaking. Joint Conference of the Metals Society and Eisenhütte Österreich, Montanuniversität Leoben, April 21–24, 1980.)
5.16. Gilli, P. V., Kemetmueller, R.: Minimizing Dust Problems in BOF Shops. Iron and Steel Engineer *1966*, 193–201.

Chapter 6

6.1. IEA Conference on New Energy Conservation Technologies and Their Commercialization (Berlin, April 1981); Berlin-Heidelberg-New York: Springer. 1981.
6.2. Uttenbroeck, J.: Seasonal Efficiency of a Heating Plant, [6.1], 329–334.
6.3. Jüttemann, H.: Elektrisch heizen und klimatisieren, 5th ed. Düsseldorf: VDI-Verlag. 1979.
6.4a. Kesselring, F., Duppenthaler, A.: The Layout of Solar Hot Water Systems, Using Statistical Meteo- and Heat Demand Data. SUN II: Proceedings of the ISES International Congress on Solar Energy (Atlanta, Georgia, 1979), Vol. 1, 884–888. New York: Pergamon Press. 1979.
6.4b. Kesselring, P.: Solar Energy and Storage, Solar Energy 1982, esa SP-181, 53–58. Paris: ESA. 1982.
6.5. Swisher, J. H., Frier, W. A.: US DOE Thermal Storage Program, [1.15], 465–500.
6.6. Zegers, P.: Overview of the Thermal Storage Work within the Energy R and D Programme of the European Community, [1.15], 443–463.
6.7. Comly, B., et al.: Ergebnisse von Forschungs- und Betriebsarbeiten auf dem Gebiete der thermischen Energiespeicherung bei General Electric, [1.16], 73–78.
6.8. Braun, J. E., et al.: Seasonal Storage of Energy in Solar Heating. Solar Energy *27*, 403–411 (1981).
6.9. Sillmann, S.: Performance and Economies of Annual Storage Solar Heating Systems. Solar Energy *27*, 513–528 (1981).
6.10. Linzer, W., Guttenbrunner, M.: Möglichkeit und Zweckmäßigkeit des Einsatzes von großen Langzeit-Wärmespeichern bei der Beheizung von Wohnbauten mit Sonnenenergie. (Fortschr.-Ber. VDI-Z., Reihe 6, Nr. 104.) Düsseldorf: VDI-Verlag. 1982.
6.11. Tabor, H.: Short and Long-Term Thermal Storage in Solar Ponds, [1.15], 179–195.
6.12. Halozan, H.: Die Außenluft-Wärmepumpenheizung – Abtauproblematik und Spitzenlastdeckung. Graz: dbv-Verlag. 1978.
6.13. Edler, A.: Verbrennungsmotor-betriebene Wärmepumpen. Graz: dbv-Verlag. 1982.
6.14. Schwindt, H. J.: Erfahrungen mit Warmwasser-Wärmepumpenanlagen sowie Aspekte aus Verbrauchersicht. Elektrowärme im Technischen Ausbau (eta) 1/2/1983; A19-A30.
6.15. Van der Ree, R.: Heat Pump Combined with Thermal Storage, [1.15], 387–414.
6.16. Holman, A. S., Brantley, V. R.: ACES Demonstration–Construction, Start-up and Performance Report, ORNL/COW-25. Springfield, Va.: NTIS. 1978.
6.17. Berntsson, Th., et al.: The Use of Ground as Heat Source for Heat Pumps in Urban Areas. Stockholm: Swedish Council for Building Research, 1980.
6.18. Brunberg, E. A.: The TEPIDUS System for Seasonal Heat Storage and for Cooling, [1.10], 247–260.

6.19. Gerken, A.: Grenzkostenbetrachtung für den wirtschaftlichen Einsatz von Wärmespeichern bei der Wärme-Kraft-Kopplung, [1.6], 153–162.
6.20. Studie über Energiespeicher in Systemen mit Wärme-Kraft-Kopplung. Studie im Auftrag des BMFT, Contract No. ET 4197A, MBB-Report No. UR-361-77 (4 Vols.). Ottobrunn: MBB. 1977.
6.21. Schäch, R.: Betrieb von Fernheiznetzen mit und ohne Speicher, Part I: Energie *25*, 157–166 (1973); Part II: Energie *25*, 216–223 (1973).
6.22. Komarov, V. N.: Selection of the Operational Mode of District Hot Water Storage Tanks (HWST) in Heating Networks. Teploenergetika *21*, 41–46 (Thermal Engineering 57–62) (1974).
6.23. Gabrysch, G.: Planung und Betrieb des 25-MW-Gasturbinenblocks im Heizkraftwerk Bielefeld. VGB Kraftwerkstechnik *60*, 475–486 (1980).
6.24. Andersson, S., Erikson, A.: Seasonal Storage in Hard Rock–Multiple Well System, [1.11], 249–258.

Chapter 7

7.1. Schüller, K. H., Haaland, M.: Pneumatisches Speicherkraftwerk ohne Brennstoffzufuhr. VGB-Tagung Kraftwerke, 1977.
7.2. Knutsen, C. A., et al.: Incremental Cost Analysis for Advanced Compressed Air Energy Storage Concepts. 14th Intersociety Energy Conversion Engineering Conference (1979), Paper 799090, 426–428.
7.3. Hamilton, N. I.: Research Studies for Adiabatic Compressed Air Storage, [1.4], 265–272.
7.4. Kartsounes, G. T., Kim, C. S.: Study of Turbomachinery Options for Compressed Air Energy Storage Plants. 14th Intersociety Energy Conversion Engineering Conference (1979), Paper 799095, 455–458.
7.5. Olsson, E. K. A.: 220-MW Air Storage Power Plant, ASME Paper 70-GT-34. New York: ASME. 1970.
7.6. Herbst, H.-Ch., Mass, P.: Das 290-MW-Luftspeicher-Gasturbinenkraftwerk Huntorf. VGB-Kraftwerkstechnik *60*, 174–187 (1980).
7.7. Vosburgh, K. G., et al.: Compressed Air Energy Storage for Electric Utility Load Leveling. 12th Intersociety Energy Conversion Engineering Conference (1977), Paper 779157, 1016–1022.
7.8. Zaugg, P.: Luftspeicherkraftwerke, Brown Boveri Mitteilungen *62*, 338–347 (1975).
7.9. Gütter, G.: Das "Pneumatische Druckspeicherwerk" – ein neuer Kraftwerkstyp zur Erzeugung elektrischer Spitzenenergie. Energietechnik *25*, 327–332 (1975).
7.10. Otis, D. R., et al.: Ejector Augmentation of the Air Supply in a Compressed Air Energy Storage Plant. 13th Intersociety Energy Conversion Engineering Conference (1978), Paper 789682, 977–983.
7.11. Lessard, R. D., et al.: Coal-fired Fluid Bed Compressed Air Energy Storage Power Plant: A Preliminary Technical Assessment. 13th Intersociety Energy Conversion Engineering Conference (1979), Paper 799089, 419–425.
7.12. Gilli, P. V., Beckmann, G.: Schnellstart von Gasturbinen mit Hilfe von VGD-Luftspeichern. Elektrizitätswirtschaft *75*, 116–118 (1976).
7.13. Gutgesell, F.: Ruths-Gefällespeicher in Hochdruck-Kraftanlagen. Allgemeine Wärmetechnik *5*, 73–81 (1954).
7.14. Schulz, E., Gropp, F.: Ruthsspeicher für Spitzenkraftwerkerzeugung in Berlin. Elektrizitätswirtschaft *29*, 153–162 (1930).
7.15. Mokesch, R.: Die 120-at Dampfspeicheranalyse der Wiener städtischen Elektrizitätswerke. Arch. für Wärmewirtschaft *19*, 87–91 (1938).
7.16. Gilli, P. R.: Sicherung und Verbilligung der Stromversorgung durch Dampfspeicher, World Power Conference (Wien, 1938), Section E, Paper 92.
7.17. Marguerre, E., Koch, J.: Gleichdruckspeicher als Ausgleich in Vorschalt- und Heizkraftanlagen. Wärme *52*, 334–339 (1929).

7.18. Oplinger, J., Anaud, R. K.: Electric Utility Evaluation of Near Term Thermal Energy Storage Systems, [1.9], 43–52.

7.19. Bechtel Corp.: Retrofitted Feedwater Heat Storage for Steam Electric Power Stations Peaking Power Engineering Study. CONS-2863-1. Springfield, Va.: NTIS. 1976.

7.20. Smidt, D.: Wirtschaftlichkeit von Wärmespeichern mit Reaktorkraftwerken. Atomwirtschaft, December 1958, 510–511.

7.21. Marguerre, Fr., Marguerre, Fe.: Anwendung der Wärmespeicherung in Kernkraftwerken und herkömmlichen Kraftwerken. Elektrizitätswirtschaft *1960*, 320–326.

7.22. Margen, P. H.: Thermal Storage in Rock Chambers. Nuclear Engineering *4*, 259–262, June 1959.

7.23. Oplatka, G.: Deckung der Lastspitzen mit thermisch gespeicherter Energie. Brown Boveri Mitteilungen *67*, 457–464 (1980).

7.24. Gilli, P. V., Beckmann, G.: The Nuclear Steam Storage Plant. An Economic Method of Peak Power Generation. Vol. IX; 162–174. Proceedings of the IXth World Energy Conference (Detroit, 1974). New York: US National Committee WEC. 1975.

7.25. Gilli, P. V., Beckmann, G.: Thermal Energy Storage for Non-fossil Power Plants, [1.8], 264–278.

7.26. Gilli, P. V., Beckmann, G.: Design and Economy of Solar Plants with Integrated Thermal Energy Storage, "Solar Energy", Proceedings of the UNESCO/WMO Symposium (Geneva, 1976); WMO-No. 477, 498–506. Geneva: WMO. 1977.

7.27. Faninger, G., et al.: Solarthermische Kraftwerke – Technische und wirtschaftliche Möglichkeiten in Österreich; ASSA FA-9. Wien: ASSA. 1979.

7.28. Sapozhnikow, F. V., et al.: Possibility of Developing Solar Power Stations in the USSR. Teploenergetika *27*, 5–8 (1980) (Thermal Engineering, 123–126).

7.29. Hallett, R. W., Gervais, R. L.: The Solar Ten Megawatt Pilot Plant, SUN II: Proceedings of the ISES International Congress on Solar Energy (Atlanta, Georgia, 1979), Vol. 2, 1137–1140. New York: Pergamon Press. 1979.

7.30. Gneuss, G.: Sonnenenergie in Kleinkraftwerken am Beispiel der Solar-Turmanlage in Almeria/Spanien. VGB-Kraftwerkstechnik *62*, 1079–1085 (1982).

7.31. The Austrian 10-kWe Solar Power Plant for Developing Countries, Final Report. Wien: Federal Press Services. 1981.

7.32. Beverly, W. D., Engle, W. W., Mahony, F. O.: Integration of High Temperature Thermal Energy Storage into a Solar Thermal Brayton Cycle Power Plant; 12th Intersociety Energy Conversion Engineering Conference (1977), Paper 779192, 1195–1220.

7.33. Ayers, D. L., Strong, R. E.: Compressed Air Storage–Another Answer to the Peaking Problem. Power Engineering, August 1975, 36–39.

Chapter 8

8.1. Partel, G.: Water and Steam Rockets. Nuclear Energy *1965*, 84–86.

8.2. Slavin, F. J.: Aircraft Steam Catapults. Mechanical Engineering *1969*, 42–46.

8.3. Foss, M.: Marine-Kunde, 7th ed. Stuttgart-Berlin-Leipzig: Union Deutsche Verlagsgesellschaft. 1890.

8.4. Müther, W.: Druckluft-Lokomotiven für Untertage-Betrieb. DEMAG-Nachrichten *1950*, 45–47.

8.5. Fauser, H.: Die Kennlinien von Druckluftlokomotiven. Glückauf *95*, 743–757 (1959).

8.6. Riedler, A.: Die Honigmannsche Dampflokomotive mit feuerlosem Natronkessel. ZVDI *27*, 729–750 (1883).

8.7. Gutermuth, M. F.: Versuche an einer Personenzuglokomotive mit Honigmannschem Natronkessel. ZVDI *28*, 533–537 (1884).

8.8. Meineke, F., Röhrs, F.: Die Dampflokomotive. Berlin-Göttingen-Heidelberg: Springer. 1949.

8.9. Gilli, P. R.: Die Höchstdruckdampf-Speicherlokomotive im Gaswerk Leopoldau. Gas/Wasser/Wärme *2*, 63–74 (1948).

8.10. Giesl-Gieslingen, A.: Über die Wahl der wirtschaftlichsten Lokomotivbauart im Verschiebe- und Zubringerdienst unter besonderer Berücksichtigung der Gilli-Hochdruck-Speicherlokomotive. (Essen, February 11, 1954.) (Heft No. 31.) Essen: Wiss. Verein für Verkehrswesen e. V. 1954.
8.11. Giesl-Gieslingen, A.: Die Hochdruck-Dampfspeicherlokomotive (Gilli-Lokomotive). Lok-Magazin (Stuttgart) *118/I-II*, 32–41 (1983).
8.12. SGP-Handbook. Wien: SGP (undated).
8.13. Stockklausner, J.: Die Wiener Lokomotivfabrik Floridsdorf. Freiburg: Eisenbahnkurier-Verlag. 1982.
8.14. "Gilli"-Lokomotive. Company Brochure, Wiener Lokomotivfabriks AG, Wien, 1940.
8.15. Edmondson, W. B.: Should We Solarize the Railroads? Solar Energy Digest *9*, 2–3 (1977).
8.16. Wollinger, A.: Der Fahrschrauber. Wochenpresse Nr. 46, 12. November 1980.
8.17. Dunn, H. S., Wojciechowski, P. H.: High-pressure Hydraulic Hybrid with Regenerative Braking. 7th Intersociety Energy Conversion Engineering Conference (1972), Paper 729145, 989–995.
8.18. Isshiki, N., et al.: Development of CDE (Concentration Difference Energy) System and Engine. 14th Intersociety Energy Conversion Engineering Conference (1979), Paper 799430, 1998–2003.
8.19. Balkan, S.: Wärmespeicherantriebe für einen schadstofffreien Fahrzeugbetrieb, ZVDI *117*, 422–429 (1975).
8.20. Holland, M. B.: The Stirling Engine–Power Plant of the Future? CME *1979*, 60–61.
8.21. Waters, E. D., et al.: The Thermal Energy Storage System for a Stirling Engine Powered Highway Vehicle, 14th Intersociety Energy Conversion Engineering Conference (1979), Paper 799098, 475–480.
8.22. Farahat, M. M.: Thermal Energy Storage Applications for Highway Vehicles with Stirling Engines, [1.4], 243–263.
8.23. Anon.: Has the Ideal Mid-temperature Thermal Storage Material Been Found? Solar Energy Digest *18*, 2–3 (1982).

Subject Index

IBM-Composersatz: Springer-Verlag Wien
Umbruch und Druck: Novographic, Ing. Wolfgang Schmid, A-1230 Wien

Energy Demand: Facts and Trends

A Comparative Analysis of Industrialized Countries

By **B. Chateau** and **B. Lapillonne,**
Institut Economique et Juridique de l'Energie, Grenoble, France

1982. 26 figures. XV, 280 pages.
ISBN 3-211-81675-5

The relationship between economic development and energy demand is investigated in this book. It gives a detailed analysis of the energy demand dynamics in industrialized countries and compares the past evolution of the driving factors behind energy demand by sector and by end-uses for the main OECD countries: residential sector (space heating, water heating, cooking ...), tertiary sector, passenger and goods transport by mode, and industry (with particular emphasis on the steel and cement industry).

This analysis leads to a more precise understanding of the long-term trends of energy demand; highlighting the influence on these trends of energy prices, especially after the oil price shocks, and of economic development patterns.

This book summarizes the authors' eight years' study of energy demand analysis and forecasting and energy conservation. It gathers together the rich supply of information — both qualitative and quantitative — they had access to.

Springer-Verlag Wien NewYork